HOW TO BUILD

CHICKEN COOPS

SECOND EDITION

HOW TO BUILD CHICKEN COOPS

EVERYTHING YOU NEED TO KNOW

PLANNING AND DESIGNING • TOOLS AND SKILLS
CARE AND MAINTENANCE • BIRD SAFETY

DANIEL JOHNSON AND SAMANTHA JOHNSON

VOYAGEUR
PRESS

For Marie, who enthusiastically shares her love of chickens with us.

Quarto.com

First published in 2015 by Voyageur Press, an imprint of The Quarto Group, 100 Cummings Center, Suite 265-D, Beverly, MA 01915, USA. T (978) 282-9590 F (978) 283-2742

10 9 8

ISBN: 978-0-7603-6411-6

Digital edition published in 2019
eISBN: 978-0-7603-6412-3

Library of Congress Cataloging-in-Publication Data

Names: Johnson, Daniel, 1984- author. | Johnson, Samantha, author.
Title: How to build chicken coops : everything you need to know / Daniel Johnson and Samantha Johnson.
Description: Second edition. | Minneapolis, MN : Voyageur Press, 2019. | Includes index.
Identifiers: LCCN 2018036247 | ISBN 9780760364116 (sc)
Subjects: LCSH: Chickens–Housing.
Classification: LCC SF494.5 .J63 2019 | DDC 636.5–dc23
LC record available at https://lccn.loc.gov/2018036247

Acquiring Editors: Elizabeth Noll, Dennis Pernu
Project Manager: Alyssa Lochner
Art Directors: Brad Springer, Cindy Samargia Laun
Cover Designer: James Kegley
Layout: Danielle Smith-Boldt

Front cover: *Top left, The Len/Shutterstock. Bottom left, Local Favorite Photography/Shutterstock. Center left, right: Daniel Johnson*
Back cover: *Daniel Johnson*

Printed in China

CONTENTS

INTRODUCTION

LET'S BUILD A CHICKEN COOP!

So you've decided that chicken keeping is something you'd like to try. You're in good company—people have been flocking (pun intended) to raising chickens in ever-increasing numbers. After all, the idea of having fresh eggs on a daily basis sounds quite inviting, and it's exciting to think that you can raise your own birds even if you don't have acres of farmland or live in a rural location. Even urban chicken keeping is a hugely popular endeavor, as it allows city dwellers to move toward self-sufficiency in a very rewarding way. Add in the benefits of chicken manure as garden fertilizer, the delight of watching a chicken's daily antics, and the satisfaction that comes from keeping livestock—even on a small scale—and it's easy to see why chickens are so popular!

Above: Whether your coop is small and charming or large and elaborate, the basic needs of your flock must be met. From choosing the right wire for the runs to making sure there's enough ventilation, there are many decisions to make when constructing your own coop.

Opposite: Raising chickens is a rewarding experience and is growing in popularity throughout the country. A hen and her chicks can provide endless hours of enjoyment for their owners. Nothing is more endearing than newly hatched chicks.

CHICKEN TRIVIA

- **How many eggs can one hen lay?** While the figure varies among breeds, the average hen can lay approximately three hundred eggs per year.
- **How many chickens are there in the world?** There are more chickens in the world than people—some sources say about 19 billion chickens worldwide.
- **Are chickens carnivores, omnivores, or herbivores?** Chickens are omnivores.
- **Is egg color related to the color of the hen's earlobes?** Generally speaking, yes; but exceptions do occur. (Did you know chickens have earlobes?)
- **Why did the chicken cross the road?** Sorry, we don't know, but we think she was in a hurry to get home to read this book!
- **Why do roosters dance?** Because they don't know how to sing. Seriously, they perform a little dance for their hens called *tidbitting.*
- **If a rooster lays an egg on the roof of the chicken coop, which direction will it fall?** Down. Ha! (Yes, we know roosters can't lay eggs.)
- **Why do chickens like to roll around in the dirt?** Chickens take dust baths to clean themselves; the dust helps to eliminate external parasites.

CHICKEN LINGO

- **Bantam:** a smaller variety of chicken
- **Broody:** the term for a hen that is interested in hatching a clutch of eggs
- **Chick:** a baby chicken
- **Cockerel:** a young male chicken
- **Coop:** the structure in which chickens are housed (but you already knew that, right?)
- **Egg:** what baby chicks hatch from (and a great breakfast food)
- **Flock:** a group of chickens
- **Free-range:** the act of allowing chickens to roam and forage at will outside of a coop or fenced-in run
- **Hen:** an adult female chicken
- **Nest box:** the portion of the coop in which the hens lay eggs
- **Pullet:** a young female chicken that hasn't started laying eggs
- **Roost:** wooden poles on which chickens can rest (also known as a *perch*)
- **Rooster:** an adult male chicken (also known as a *cock*)
- **Run:** a large enclosed area where chickens can exercise

But if you want to keep chickens, they're going to need a place to live, and that's why you're holding this book in your hands. Providing your chickens with a safe, comfortable habitat is your first priority as a chicken owner. Ideally, you'd like to outfit your birds with a home that will meet all of their needs as well as prove to be an attractive element to your yard or farm. But maybe you're on a budget and want to keep your costs down. Maybe you like the idea of customizing your coop, or maybe you have materials and supplies on hand and would like to be able to incorporate those into your coop. Or maybe you have an existing structure—such as a playhouse—that you'd like to convert into a chicken coop. As with so many farming endeavors, there are as many answers as there are questions when it comes to raising chickens. People have different experiences and varying ideas about the best ways to raise and house chickens, so we'll present you with a lot of questions and hopefully as many answers.

Chickens are a great family project, and building a coop together can be very satisfying.

A good coop is a safe and happy place for your chickens. They'll willingly enter their coop to roost at the end of the day, keeping them safe from predators.

SAFETY FIRST!

Always exercise caution when beginning a construction project. Safety must be your number one priority when working on your coop. It is vital to be aware of your tasks and surroundings and pay attention to tool and construction safety. Learn to use your tools in the safest and most effective ways, and avoid placing yourself in any situation that could prove dangerous. Remember, many of the activities described in this book can be dangerous. Failure to follow safety procedures may result in serious injury or death. The authors cannot assume responsibility for any damage to property or injury to persons as a result of the use or misuse of the information provided.

Collecting fresh eggs for your family and enjoying their wholesomeness is a big reason that people start raising chickens. Some also start a small egg business if their chickens begin producing more than they can use.

There are a lot of possibilities to think over when deciding the size and type of coop you'd like to have for your own flock, and that's why we wrote this book: to help you navigate the waters of coop building in an effective and straightforward way. Whether you have extensive carpentry experience or are a newbie with a hammer and nails, we'll take you step-by-step through the process of creating a coop that will be your flock's very own "coop sweet coop."

Let's get started!

Even the youngest members of the family will enjoy getting to know the chickens and will learn about their care and feeding.

1 WHY YOU NEED A COOP

Shelter is one of the most important aspects of providing a coop for your chickens. Chickens benefit from a shady location if you happen to live in an area where temperatures soar during the summer.

In this chapter, we'll discuss a major question: Why do you need a chicken coop? While it's likely that you already have a reason or two in mind, we'll present a few more thoughts and ideas, offer some advice, and point out some of the most exciting advantages of embarking on this type of project.

The primary purpose of a chicken coop is, of course, to protect your chickens from adverse weather and predators. Secondarily, a coop provides a specific place for your hens to lay their eggs. The average prefabricated coop will address these needs, but building your own coop gives the opportunity to save money, customize your design, and have a lot of fun in the process. But before we explore the benefits of building your own coop, let's jump in and learn more about why chickens need coops.

SHELTER

Chickens require sturdy, well-ventilated housing to avoid the discomfort and health concerns that result from exposure to the elements. While a little bit of rain generally won't bother your birds, heavy rains—especially cold rains—are another thing entirely. Most chickens don't mind a light drizzle or a passing shower, thanks to the feathers on most chicken breeds that repel water quite effectively. But you wouldn't want your feathered friends to be forced to weather a downpour or a snowstorm, and a sturdy, dry coop provides precisely

The importance of good housing for your chickens can't be overemphasized. The health and safety of your flock should be your prime concern. Most chickens enjoy free ranging outside of their coop for periods of time during the day. You'll need to do what you can to keep them safe.

the protection your birds need to avoid unpleasant weather.

For cold climates, a coop is essential for your chickens during winter. While some breeds are quite hardy and don't seem to be bothered by temperatures that are less than balmy, cold winds can be very hard on livestock of all kinds, including chickens. The same is true in the case of snow; most chickens can handle snow in moderation, but it is very important for them to have a dry location where they can get out of the cold and wet conditions. Additionally, chickens are susceptible to frostbite on their wattles and combs, and their feet may also be at risk if they stand for too long in deep snow. A chicken coop provides a cozy place for chickens to escape the elements.

In warmer seasons, a coop can provide some much-welcomed shade for your chickens, but if you live in an area with summers that are routinely quite hot, you'll want to consider modifications to your coop design to ensure that your birds don't get overheated while in the coop. These options could include increased ventilation, window shades, and provisions for the temporary use of electric fans during hot weather. Again, these modifications won't be necessary for all chicken owners, but in very warm climates they may be an essential step in maintaining happy, healthy birds.

PREDATORS

Protection from predators is just as important to your chickens as shelter is—perhaps more so. While you would never let a fox guard your henhouse, will one be able to sneak in? Your coop and its run (and any additional fencing) act as your chickens' primary protection from sneaky critters who would like nothing better than to grab a quick chicken snack. Let's take a look at some of these potential predators, and then some ways to prevent their access to the coop.

PREDATORY BIRDS

It might seem surprising, but other birds are among the greatest threats to your chickens. It's not unheard of for predatory birds such as

A rigorous climate—such as those found in the northern parts of the country—means that you'll need to provide a coop that enables your birds to escape from the snow and bitterly cold temperatures.

Hawks are a common predator of chickens, and they'll think nothing of stealing your hens. Maintaining a good roof on your coop and runs will deter most large bird predators.

owls, eagles, and hawks (including the red-tailed or "chicken" hawk) to swoop in and make off with a chicken or two. These predatory birds are very strong and brilliant hunters—don't underestimate them. Your chickens are at the highest risk of a bird attack if they are allowed to free range, and chickens safely tucked in a coop with an attached run will face far fewer dangers from other birds.

CANINE THREATS

Foxes, coyotes, and even domestic dogs can pose a significant threat to your flock. Depending on the time of year, foxes and coyotes may be feeding their pups, and they're often on the lookout for additional food sources. On the domestic side, an overexuberant pet dog may spell big trouble for your chickens at any time of the year. Even a dog that is otherwise trustworthy around the farm may become a chicken chaser if the opportunity arises. Dogs from neighboring homes or farms can be among the worst threats to your birds, so always be vigilant about keeping your chickens in the coop after dark.

CATS

Chicken attacks from wild cats (such as bobcats or mountain lions) are not as common as attacks from canines, but they do happen. Surprisingly, domestic cats can also be a source of trouble; some cats simply enjoy the hunting process and will stalk other animals (including chickens)

Your neighbor's dog might be your flock's worst enemy. Dogs love to chase anything that moves and even your own dog might make the mistake of chasing one of your hens. Be sure to train your dog to leave your hens alone.

Although you might not think about cats bothering your chickens, they certainly can be a problem. (But usually they are helpful in rodent control around your coop and yard.)

PROTECT YOURSELF

While you might think it makes sense to take a defensive strategy against predators that threaten your chicken coop, think twice before you shoot, trap, or relocate those troublesome individuals. States, cities, and towns have varying laws regarding the protection of predator birds and animals, and unless you are familiar with the laws and ordinances applicable in your community, you could inadvertently and unknowingly break the law.

On the federal level, the Migratory Bird Act protects many predatory birds (including hawks), making it illegal to "pursue, hunt, take, capture, kill, attempt to take, capture or kill, possess, offer for sale, sell, offer to purchase, purchase, deliver for shipment, ship, cause to be shipped, deliver for transportation, transport, cause to be transported, carry, or cause to be carried by any means whatever, receive for shipment, transportation or carriage, or export, at any time, or in any manner, any migratory bird, included in the terms of this Convention . . . for the protection of migratory birds . . . or any part, nest, or egg of any such bird."

There may be local restrictions regarding other predators, such as raccoons. Some cities prohibit the use of firearms entirely, while other locations may require the possession of a permit before shooting an animal. The use of a live trap and subsequent release of the animal in a new location can be an option in certain areas, but again—get the facts first.

The bottom line is, educate yourself before trying to eradicate predator birds and animals. You do not want to find yourself facing a large fine (or even jail time!) for killing a protected predator.

for sport. While it might be uncommon for a domestic cat to kill an adult chicken, cats are often more than happy to go after chicks or small bantam breeds. The advantage to having a domestic cat around your chicken coop is the cat's ability to hunt and eliminate rodents—such as mice or rats—that could be attracted to the feed around your coop.

WEASELS, SKUNKS, AND RACCOONS

Even small animals can be troublesome, so be aware of the hunting abilities of sneaky critters such as weasels, skunks, and raccoons. All three are capable of squeezing, burrowing, climbing, and digging their way inside your coop. Weasels can squeeze through very tiny spaces and are unfortunately quite good at chewing the heads

Raccoons are some of the cleverest adversaries you'll meet when it comes to keeping chickens. They are capable of opening latches and squeezing into surprisingly small spaces.

off of chickens. Raccoons are smart, persistent, and very strong for their size, so they can be one of the biggest challenges for the chicken owner; raccoons are also experts at opening latches and gates to get at what they want—so be on your toes.

RATS, OPOSSUMS, AND MINKS

While it is possible for these smaller predators to take down an adult chicken (especially smaller bantams), you may find that rats, opossums, and minks are more interested in trying to steal your eggs or chicken feed. Minks are semiaquatic and require a nearby water source, so if you live far from a lake, river, or pond, you probably won't be seeing these pests, but they are considered to be one of the more notorious chicken killers; keep an eye out for them. Rats and opossums are only prevalent in certain parts of the country, so you may not have to deal with them in your area, but rats can be hard to eliminate once they become entrenched in a certain location.

PREDATOR PREVENTION

So, what are the best ways to deal with a predator problem? While it's virtually impossible for you to maintain an environment that is 100-percent safe for your chickens all the time, you can take specific steps to lower or eliminate

Cooping your chickens at night will ensure their safety, as most predators that hunt are nocturnal.

the danger of predator attack. These include the following:

- **PUT YOUR CHICKENS IN AT NIGHT.** Many predators hunt at night; it only makes sense to protect your flock by putting the coop to use during the nocturnal hours. Even if free ranging is important to you, allowing your chickens to go unprotected through the night will almost always lead to trouble.
- **BURY THE FENCE SEVERAL INCHES.** To prevent animals from gaining unwanted access to your coop by digging under the run, bury the wire at least several inches underground; this can help keep out domestic dogs, foxes, coyotes, and even small animals such as weasels and opossums.
- **USE STRONG MESH FOR FENCING—NOT CHICKEN WIRE.** Chicken wire is best used as a method of controlling and maneuvering the chickens themselves, not as the prime barrier between the flock and the predators. Larger predators can tear chicken wire and rip it down; smaller predators (especially rats) can "weasel" their way through small openings. Instead, opt for stronger wire mesh, such as hardware cloth, with smaller openings.
- **KEEP LATCHES LOCKED.** Some predators—raccoons in particular—are extremely clever about opening gate latches, or working a gate until it gives way. For these reasons, gate integrity is a prime issue for the chicken owner. In addition to installing strong hinges, make sure that the gate fits snugly into the frame and that there are no gaps. But perhaps most important of all is to be sure that the gate latches are locked in a critter-proof manner. Snaps and padlocks—even something as simple as a piece of twisted wire!—can go a long way toward preventing your lovely chickens from falling into the paws of a greedy raccoon. Once you have a gate with a sturdy lock, take steps to establish an evening coop-locking schedule and chicken head count—doing this at a regular time each night will help ensure that the coop is not inadvertently left open overnight.
- **GET A GUARD DOG.** While it's true that your chicken coop is your number one defense against predators, a guard dog

Burying wire is one of the best ways to deter predators from accessing the interior of your coop or runs, but another terrific idea is to place your coop over brick pavers at a width of about 18 inches. When the predator tries to dig, they won't find easy access and will typically give up.

This Golden Laced Polish resides inside a poorly constructed chicken run. The use of chicken wire rather than hardware cloth and the large gaps on the bottom leave this chicken vulnerable to attacks from predators.

Livestock guardian dogs such as this Great Pyrenees provide protection and added safety for your chickens and other livestock.

that is properly trained can be a great help in deterring predators as well. If your dog has been introduced to the idea that chickens are not playthings but rather something in need of protection, it can be another solution for preventing attacks from ground-based animals such as foxes, coyotes, raccoons, weasels, and skunks. In many cases, simply the presence of a dog on your property will be enough to discourage such predators. Be aware, however, that you do not want your dog to actually tangle with any of these animals—in addition to the injuries your dog could incur, wild animals potentially carry diseases, such as rabies. The intimidating presence of the dog—not physical contact with the predators—is what you're after.

- **DISCOURAGE PREDATORY BIRDS WITH SHINY OBJECTS.** Some people have had good success deterring eagles and hawks by hanging and arranging shiny objects (have any old CDs around?) on and near the coop. The predators don't seem to like the objects and it certainly has the advantage of being an inexpensive option.
- **COVER THE TOP OF YOUR RUN WITH CHICKEN WIRE/MESH.** If predatory birds are an issue in your area, you may want to consider protecting your chickens from aerial attacks by enclosing the top of their run with some sort of wire mesh.
- **AVOID LEAVING PREDATOR-ATTRACTING ITEMS LYING AROUND.** Easily accessible garbage, pet food, and even potential critter housing (piles of wood or other materials) should be avoided whenever possible, as these items attract wild animals. Even small predators can cause headaches for you and your chickens, so try not to inadvertently encourage their presence.
- **SET TRAPS.** Rats, of course, can be trapped with rattraps, and some of the other smaller predators (raccoons, opossums, etc.) can be trapped live and perhaps relocated (see the "Protect Yourself" on page 18).

ADVANTAGES TO BUILDING YOUR OWN COOP

While you could go down to the nearest farm supply store, pick out a prefabricated chicken coop, and set up a home for your chickens in a single afternoon, building your own coop from scratch has distinct advantages. You'll have the opportunity to advance your carpentry skills while constructing a coop that will become an attractive component of your current landscape. Additionally, building your own chicken coop

Designing and building your own coop allows you to add as many extras as you wish. Nice accents to the door and exterior access to nest boxes are just two of the embellishments you can consider.

BEAUTIFICATION: YOUR COOP AS PART OF THE LANDSCAPE DÉCOR

What's prettier than a colorful flock of hens against a picturesque backdrop of grass on a beautiful sunny day? Don't let your chicken coop detract from this overall picture of avian bliss. After all, a chicken coop doesn't have to be strictly utilitarian—it can also be an important element in the existing landscape of your yard or farm.

When planning the layout and design of your coop, consider the overall aesthetics of your home and current outbuildings, then tailor the design of your chicken coop to accentuate and elevate these elements. By incorporating your chicken coop as a cohesive piece of your landscape, you'll increase the overall appeal of your surroundings. And by building your own coop, you can pursue these beautification ideas without being restricted by a certain design, layout, or look.

This is a lovely setup for raising chickens. Two portable coops placed end to end allow space to raise more birds, including those of varying ages. This coop adds to the beauty of the yard and is a welcome addition.

A chicken coop can range from fairly easy to very complex, but it can be a fun and challenging project to tackle.

presents many other advantages—here are some of them:

COST SAVINGS

Depending on the size, scope, and complexity of the coop that you choose to build, your potential costs will vary considerably. By choosing to build the coop yourself (as opposed to buying a prefabricated model or hiring a contractor to handle the construction), your expenses can be lower, representing a significant cost savings over the course of the project. Can you fit the project in on weekends or in the evenings? Can you round up a few friends who are handy with tools and have a penchant for poultry? Will you have to buy all of the materials, or do you have scrap lumber and wire on hand? If you can utilize and repurpose materials that you already own, you can save money while constructing a quality chicken coop that will be suitable for your flock.

CUSTOMIZATION OPTIONS

The sky's the limit when it comes to possible design options because chicken coop styles are as wide-ranging as home designs. Spend a few minutes browsing the Internet (try Pinterest or houzz.com for a great selection of ideas), and you'll be pleasantly overwhelmed by the vast array of options that exist for chicken coop design. It's easy to incorporate your favorite elements and make adjustments to meet your needs.

Do you want a chicken coop that gives the appearance of a small house? Do you prefer a coop that sits on top of a run? Do you want a coop that is simple and practical, or one that takes a more whimsical approach? By constructing your own coop, you'll be able to customize its elements to suit your specific needs and preferences, as well as the size of your flock and your plans for future expansion.

FUN CHALLENGE

Okay, so you probably wouldn't go out and build yourself a house with nothing but a set of plans and some scrap lumber (or maybe you would—if so, congratulations on your skills and ability!), but you don't have to let a lack of previous construction experience stop you from building your own chicken coop. Think of the project as a fun, small-scale building challenge that will help you hone your hands-on skills while offering many benefits and very few pitfalls. What better way to learn basic building skills than to practice on your very own chicken coop? After all, even if you make a minor mistake, the solution is likely a simple fix. And if your coop isn't 100-percent perfect in every way, your chickens will probably withhold judgment and put the coop to good use anyway.

FAMILY PROJECT

What better way to embark on a fun family project than to work together and build a coop? Use it as an opportunity to learn the basics of construction as you share memorable moments and prepare an accommodating abode for your chicken residents. Everyone in the family will learn practical skills involving tools and measurements, as well as receive practice in following directions and making adjustments. Your family will learn the necessary steps to provide a safe and protected haven for your chickens, and important aspects of livestock husbandry. By involving the entire family in the coop-building project, you'll not only build important lifelong skills, you'll develop a feeling of connectivity and togetherness in your new chicken-keeping endeavor.

Working together on a family project—such as a chicken coop—offers a lot of perks, such as spending time together, practicing new skills, and learning how to handle tools safely.

2 CONSIDERATIONS BEFORE BUILDING

Before you gather up that scrap lumber and tie on your tool apron, you'll want to consider a few things. Where are you going to build this chicken coop? Have you thoroughly evaluated the available locations and chosen a site that will be the most effective place for your coop? Have you considered the presence of sun, shade, and wind? Have you considered legalities and ordinances that might be in effect? Have you figured some ballpark costs? Have you figured out where you'll buy your chickens? And what breed(s) will you choose?

The best time to ask these questions is before you start sketching plans or buying lumber and certainly before you begin to build. So try to establish a clear picture in your mind of exactly what your chicken coop project is going to entail. In this chapter, we'll walk you through these steps so you can proceed with your project confidently and with the knowledge that you've carefully considered your options and made smart decisions. Let's get going!

COSTS

Okay, admit it, this is the question that you're wondering about most of all. Just how much is this chicken coop going to cost, anyway? It's a legitimate question, and one that is wise to consider before you begin building an elaborate chicken mansion that you later discover is far out of your price range. At the very least, you need a ballpark figure on which to build.

There's only one thing: coming up with that ballpark figure isn't necessarily easy. The number of variables involved with this type of project can make it difficult to establish a basic cost. What kind of variables, you ask? Here are a few:

- **THE PRICE OF LUMBER:** This varies regionally, seasonally, and by type. Pine is generally cheaper; cedar is more expensive. The type of lumber you choose will have a direct and significant effect on the overall cost of your coop.
- **YOUR SKILL LEVEL:** Can you do most of the work yourself? Or will you need to hire someone to handle parts of the project? Do you have friends or relatives with the skills to help you construct your coop?
- **THE SIZE OF YOUR COOP AND ITS DESIGN:** Will you choose a metal roof? Or cedar shingles? Do you need a coop that is large enough to house a dozen hens, or will a four-hen coop suffice? Is the coop going to be moveable or stationary? Your choices of material and the finished size will have a direct impact on the cost of your coop.
- **EXISTING MATERIALS AND SUPPLIES:** Do you have leftover scrap lumber from a previous project that you can repurpose for your chicken coop? Or will you have to invest in new lumber for the entire project? Hardware makes up less of the cost than lumber but should also be figured in.
- **EXISTING TOOLS:** Is your garage already equipped with a chop saw and a compressor? How about a cordless drill? If you already have access to these vital tools,

You'll need to ask yourself a few questions prior to purchasing a flock of chickens or ordering chicks. Where will you build your coop, how much do you want to spend, and do you have the necessary skills to accomplish the construction of the coop yourself?

CHOOSING LUMBER

Whenever you're selecting lumber for a project, but especially when you're building something like a chicken coop that needs to have form as well as function, it's wise to choose carefully. Don't just grab the first boards on the stack at the lumber yard; instead, do what the pros do: make a brief examination of key lumber pieces. What are you looking for? Consider these points:

Knots. Knots are the brown spots on a piece of lumber where a branch was growing while the tree was alive. In some cases, knotty wood is desirable for aesthetic reasons (for use in home paneling, for example), but for woodworking, it can be problematic. Knots are incredibly hard, which makes them challenging to nail, screw, or cut through, so it's best if you forego lumber with a large number of knots (some knots are, of course, inevitable). You can also look for lumber that will work for you once you cut the knots off.

Straightness. Ideally, every piece of lumber you choose will be straight, but wood is a natural material, so some variations are to be expected. Depending on the particular piece (and, to some extent, how it was stored), you may find some warping along the edges. When choosing lumber to purchase, "sight" each piece by looking down the long end. If it's fairly straight, then you should be good to go, but if there are significant bends or obvious twists in the board, you might put it back on the shelf and try another.

Crowns. For wider pieces of lumber such as 2 × 10s and 2 × 12s, make a quick check for a "crown" when you're sighting the straightness of each board. Different from bowing along the length of a board, a crown is an arching of the short end, also called "cupping." It is usually okay if a board has a slight crown; in fact, it can be used to add strength to the project. For instance, if any of the six 8' 2 × 12s used on the roof of the coop in this project has a slight crown, the arch of the crown should be placed "up," since it will then help give the board a little more strength against gravity and its heavy load of shingles.

Take your time when shopping for lumber. Wood is a natural product with inevitable accompanying blemishes and defects, so it's smart to conduct a brief inspection before purchase.

you'll have fewer immediate expenses than someone who must go out and purchase tools before starting on the project. Or is it possible for you to borrow some of the more expensive tools from a family member or neighbor? If not, and if the cost of purchasing tools is prohibitive, research tool libraries in your area (www.localtools.org/find); these allow you to access most tools for free or for a small membership fee. You can also inquire whether your hardware or home improvement store will allow you to lease power tools by the hour or day.

- **STARTING FROM SCRATCH OR REPURPOSING ANOTHER BUILDING**: Perhaps you have a child's playhouse that's no longer being used, an extra shed on your property, or an old vehicle that no longer serves its original purpose. Or do you like the idea of designing and building your own coop? Starting with a preexisting building will save some dollars, but it may limit some of your other design options.

So while one person might be able to build an entire coop for $600, someone else might build an identical coop at a cost of $1,000. Get out your pencil and paper, and maybe a calculator, and figure out what it might cost you before you begin building.

CHOOSING A LOCATION FOR YOUR COOP

Having made the decision to build your own chicken coop, one of the first things to consider—and indeed, one of the most important things to consider—is where it should be built. There are many factors to think about when choosing the location for your chickens' happy home. Making the right decision is very important for the wellbeing of your chickens, as well as to save you from unnecessary expenditures of time and effort.

After all, you don't want to go to all the trouble of building the world's greatest chicken coop only to realize that a poor choice of location has introduced several unavoidable problems that will seriously hinder your quest to raise poultry. And as you will soon discover, once a chicken coop is fully constructed, transporting it to a different location is not always an easy task!

Of course, you could consider a mobile chicken coop, sometimes called a chicken tractor. If you have the room to move your coop from place to place on your property, this is an attractive option. At this point we're going to be discussing the coop as a permanent fixture and not a moveable coop; we'll go into that possibility on page 62.

GROUND CONDITIONS

To prevent yourself from traveling down blind alleys or embarking on wild goose (or chicken!) chases, devote a good deal of time to choosing the ideal location for your coop. One of the first things to look for is the condition of the ground. It is not advisable to build your coop on a moist site, as wet conditions promote the development of mold. If your coop is made of untreated wood, moist ground can lead to the

Preparing a site for your coop might entail some groundwork if your coop is going to be a permanent structure.

decay of your coop due to rot. Additionally, there are many health issues to consider, and a wet or damp site is not ideal for your chickens' health. So be sure to choose a dry site that will remain free of standing water or general wetness, even after a heavy rain.

In examining ground conditions, you should also look for a level site without slopes (unless your coop is to feature a professionally built, level foundation) because building on a sloped surface will not only make caring for your chickens difficult, but it could introduce the possibility of the coop tipping over in the right circumstances. A slight slope can be corrected by propping up the low side of your coop with a sturdy foundation of concrete blocks, but if the slope seems too steep, you might want to choose a different place.

SEPARATION VS CONVENIENCE: THE GREAT DEBATE

One of your primary considerations when choosing the location for your coop is whether you prefer the convenience of having the coop located close to your house or the aesthetic and aromatic benefits of having the coop located further away. If you live on a small lot, you may not have much choice, but if you live on a larger parcel, you will soon discover that choosing between separation and convenience is a multifaceted decision without a clear-cut answer.

Do you want your coop placed in close proximity to your home for the added convenience, or would you prefer to have it a bit further away? If your yard is large enough to give you a choice in location, consider all the variables prior to building.

There are obvious benefits and drawbacks to both possibilities. If you shudder at the thought of awakening every morning to the sight of a chicken coop right outside your door and if the arrival of "fowl" odors every time you open your windows is not something you desire, then locating your coop farther away from your house could be a relief to you and your family. Then again, if occasional odors don't bother you, and if the sight of your coop and your chickens gives you a sense of satisfaction and accomplishment, then positioning your coop closer to your house will allow you to reap the rewards that a more distantly placed coop cannot offer. We should mention, though, that a clean and well-maintained coop offers little odor, and an attractive coop can be a pleasant addition to your yard.

One of the primary benefits to keeping your coop near your home is its proximity to water and electricity. The farther your coop is from a water source, the farther you will have to haul fresh water for your chickens—a chore that will need to be done frequently. Therefore, regardless of whether you locate your coop close to or far away from your house, you do not want to place it at the top of a steep hill several hundred feet from the nearest water faucet. Even the most devoted chicken enthusiasts may find the trek tiresome after a few days. Locating your coop close to your house will shorten the journey from water source to coop, saving you time and effort in the long run.

Along similar lines, placing your coop close to a source of electricity will enable you to more easily include lights and/or a heated water supply in your coop, which are especially helpful if you live in a location with cold winter months. Even if you won't need electricity, you might still want to build your coop near a power source, if not just for the convenience of being able to use corded power tools while building the coop.

If your coop is less than 100 feet away from the nearest electrical outlet, a good extension cord may be sufficient to occasionally power a light. However, you should consider some safety concerns, such as the power demands of your equipment (does your equipment need more electricity than the extension cord can provide?) and the possibility of small rodents (such as mice or squirrels) chewing through the cord. For longer distances—or for peace of mind when powering electricity-guzzling equipment—it might be wise to have a professional electrician install permanent electrical wiring in the form of a buried or overhead cable to bring electricity directly to your coop. However, the farther your coop is from an electrical source, the farther you will have to run the wiring, and the more expensive your project will become.

A solar power panel can eliminate the need for direct access to electricity and can be used to power things such as an electric fence for a run.

If you'll be using a chicken tractor, or moveable coop, then you can look into solar-powered units to produce enough electricity for the runs, lights, heaters, and electronic door openers. Of course, it's possible to raise chickens in a coop without the benefit of electricity for lights or heat, but when you want eggs all winter long, you'll need to provide a light source for fifteen to sixteen hours a day to encourage egg production.

Choosing a cool and shady location can be beneficial on the hottest days of summer. Your chickens will appreciate the ability to get out of the sun.

The pros and cons of separation versus convenience are numerous, and there is no right or wrong decision. Analyze the possibilities and then choose the option that makes the most sense for your situation.

SUN, SHADE, AND WIND

It is also important to consider the location of your coop with regard to the sun. In general, your coop should be positioned so that the front receives early-morning sunshine, which provides drying and warming benefits, but also receives some shade from the strong heat rays of the afternoon. Due to their lack of sweat glands, chickens do not tolerate heat very well and can suffer heat stroke if they get too hot. Therefore, making sure that your coop is protected from the sun's hottest rays will go a long way toward keeping your chickens happy and healthy.

On the other hand, if your live in a colder region of the world, your coop should be positioned to face south, guaranteeing that it receives as much warmth from the sun as possible. But regardless of whether your live in a warm or cold climate, make sure that your chicken coop's run offers shade at all times of day, so that your chickens will be able to find relief from the sun if they so desire. Trees, bushes, or shrubs can do the job well, but if such plants are not to be had, then you can construct a shady region of your run out of wood, concrete blocks, or any other sturdy, weather-resistant materials you can find.

On a similar note, your run should offer an ample supply of grass for your chickens to consume. Not only is grass an important part of your chicken's diet, but it also harbors small insects and various seeds that are another major component of your feathered friends' diets. Providing your chickens with an opportunity to munch on these goodies will help ensure that your chickens are receiving a well-balanced menu with all of the nutritional components that they require. Many people free range their flocks and only coop the chickens at night or in inclement weather. In these circumstances, the birds are allowed free access to grass and other vegetation along with all the benefits of being able to feast on lots of bugs and insects. In the case of a permanent coop, free ranging during the day is a good choice in many cases. If you don't feel comfortable letting your chickens run free during the day, consider a moveable coop or chicken tractor that can to travel to the grass whenever necessary.

Wind and ventilation are other important considerations. Your coop needs windows or vents to allow the circulation of fresh air and to thwart the possibility of respiratory diseases, as well as to eliminate odors and gases (such as ammonia). Unfortunately, like some people, chickens are not very fond of wind, and it's best to make certain that they receive protection from cold drafts. One way to do this is to position your coop so that the windows are opposite of the typical wind direction in your area. For example, many winds in the Northern Hemisphere blow from west to east, so positioning the windows of your coop on the east side can help shelter your chickens from the bitterest of winds. In addition, you can build your coop in an area where natural or man-made windbreaks—such as a line of trees, a large building, or other such obstacles—preclude the wind from reaching it.

Grass is a delight to chickens and being able to provide them with grass in their run is a real treat. In addition, if you choose to allow your birds to free range during the day, they'll likely find some grassy areas to enjoy, such as this lovely yard.

Ventilation in your coop is very important. A window that opens and closes is a nice addition to your coop design.

One final aspect to consider when choosing a site for your coop is the possibility of having a compost pile. If you intend to save your chickens' droppings for future use as fertilizer—an excellent idea if your farming endeavors extend to gardening as well—then it makes sense to site your coop near your compost pile. This will save you the great deal of time and effort of hauling the droppings long distances. Chickens also enjoy getting into the compost pile and do a good job of helping to keep it turned.

In conclusion, choosing the location of your chicken coop is an important step in your chicken-raising venture. But while finding a location with every desirable feature is ideal, the reality is that your property may not offer every possible attribute. In that case, do your best to find the ideal location, and make do as well as you can.

WHERE TO BUILD YOUR COOP

You may be thinking, *Wait a minute—didn't I just read a whole section about where to build my coop?* The answer is yes, you did, but in this case, we're not talking about where to *locate* your coop—we're talking about where you should embark on the process of *constructing* the coop.

It may seem logical to build the coop in your garage, where supplies, tools, and climate control are easily accessible. (It's not much fun building a chicken coop out in the sun on a hot summer day.) But in practice, this is not a good idea. Unless you are building a portable coop or an absolutely barebones one with thin walls and few-to-no frills, you will soon discover that your fully assembled coop is much too heavy to be easily moved.

So to avoid back-breaking effort down the road, go about the actual assembly of your coop

If you can position your coop close to your compost pile, your chickens will do an excellent job of turning the pile and keeping it aerated.

Letting your chickens free range can have its drawbacks too, especially if they decide to spend time at your front door and make a mess of your entry. You might consider building a larger chicken run or creating a yard that's fenced in to allow the chickens to roam while keeping them off your porch.

on the site where you wish for it to be located. If this means hauling tools and supplies to the location, so be it. If it means working in the sun on a hot day, put on sunscreen or set up a shade-creating tent. Believe us when we say it will be worth it in the long run, even if it doesn't seem so at the time.

HEY, DO I NEED A PERMIT? THE LEGALITY OF COOP BUILDING

When you first think about building a coop, you probably aren't thinking in terms of regulations and ordinances. After all, it's just a coop, right? But you might be surprised at the amount of red tape involved with your project. Chicken coops don't really seem like the type of structure that would require a permit for construction, but the truth is that some locations have very specific ordinances and regulations that could come into play.

But before you even get to the legalities of coop building, you'll need to first determine whether chickens may be kept in your location. Unfortunately, many cities and towns have restrictions on chicken keeping, including regulations on the number of chickens that can be kept, the size of the property on which the chickens are housed, and other criteria. Some areas prohibit the keeping of roosters due to the extra noise; others require that all chickens be leg-banded for identification purposes. Even though many locations have adopted chicken-friendly policies, you may find that your municipality is quite restrictive with regard to the keeping of poultry. Always be sure that you are in accordance with local ordinances before embarking on your chicken-keeping endeavor.

In some towns and cities, roosters are not allowed and the ordinances are very specific as to the number of birds you may house on your property. Check your local ordinances to see if there are any that apply to raising chickens.

Some areas require a permit to keep chickens, and this permit may need to be renewed on a regular basis. Additionally, some areas require that you obtain written consent from the owners of the properties that adjoin yours, giving their approval for your chicken keeping.

If you've determined that your location is zoned for chickens (yay!), then your next step will be to determine whether you will need another kind of permit: a building permit for your coop. The relatively small size of most coops alone puts them under the size criteria that requires a permit, but coops that are portable (that is, not on a foundation) are often exempt from permit requirements altogether. Check with your local zoning department for information on specific legalities for your area. For example, if your local regulations stipulate that structures over 10 × 10 feet require a permit, and your chicken coop is 4 × 8 feet, then your coop doesn't require a permit. However, if your coop is going to be 10 × 12 feet, then you would need to pull a permit.

After you've researched and obtained any necessary permits, you'll need to ensure that the plan or design for your chicken coop meets the criteria for chicken housing as outlined in your local regulations. You might be surprised at the detailed requirements that some locales have enacted with regard to chicken coops. Some ordinances detail the size required for a coop, the number of windows that must be included, the minimum height of the structure, the number of doors, and other criteria. Similar regulations might also apply to enclosed chicken runs; this portion of your structure may need to be a specific size and height. There are often additional restrictions with regard to chicken butchering, the storage of chicken feed, the regular disposal of manure, and other areas of maintenance.

Think you have everything covered? Not so fast. Ordinances may regulate the location of a chicken coop on your property. Many stipulate a specific "setback" that defines how far away your coop must be from property lines, the street, and so on. The size of your parcel of land can play an important part in this; larger lots or acreage parcels obviously offer more options for coop placement, whereas small urban lots may have limitations in finding an appropriate and legal place for a coop.

Hatching chicks is a rewarding and fascinating adventure.

Do you need to know where to find ordinances for your area? Contact your local courthouse, or do a quick web search for "keeping chickens, ordinances, [your town and state]". Your County Extension Service agent may also have helpful information. By carefully following these regulations, you'll ensure that the coop you build is in compliance with local ordinances and that your chickens will be a welcome (and legal!) addition to your community. If you find that you are located in a less-than-chicken-friendly town or city, don't be too quick to despair—many towns are reconsidering the restrictiveness of their ordinances and codes and may very well be open to discussion on changing some overly heavy-handed regulations that exist.

GETTING STARTED WITH CHICKENS

Getting started with chicks is a rewarding experience, but it can be a little confusing if you've never done it before. Before you purchase your chicks, your first step is to settle upon the breed (or breeds) of chicken that you'd like to keep. Your decision will depend on a variety of criteria, from size to egg production, and it's important that you put careful thought into your birds before you buy. There are many websites and books devoted to chicken breeds and their various attributes. The information is voluminous and not within the scope of this book, but there's loads out there, and talking to other chicken enthusiasts is a great start.

If you're keeping chickens with an eye toward those farm-fresh eggs, then you might want to consider one of the well-known egg-producing breeds, such as a White Leghorn. Breeds that are noted for their impressive egg-laying ability might produce as many as 300 eggs per year, while other breeds might produce only 100. So if eggs are your aim, bear this in mind when selecting a breed.

If you'd like to raise chickens as meat birds, then you're looking for an entirely different set of characteristics. In this case, you might want to try a hybrid, such as the Cornish Rock Cross, which is noted for its ability to gain weight quickly.

If eggs *and* meat are important to you, then a dual-purpose breed—perhaps the Rhode Island Red—could be a good choice. Dual-purpose breeds may not achieve the impressive production rate of specialty breeds, but they provide a great option if you're looking to keep only a few birds and want to diversify your focus.

Some people find the idea of raising endangered chicken breeds appealing. Check out The Livestock Conservancy (formerly the American Livestock Breeds Conservancy, www.livestockconservancy.org) for a list of rare-breed chickens. A few breeds on the "critical" list include the Russian Orloff, the Nankin, and the Spanish.

You may also want to consider other aspects, such as the hardiness level of the various breeds—their ability to withstand extremes of temperature. (Do you live in a frigidly cold or swelteringly hot climate?) Find out if your breed of choice will be happy and healthy in your climate before you buy.

Once you've selected the breed (or breeds) of chickens that you would like to raise, the next step is to figure out where to purchase your poultry. One popular and easy option is to buy them through a mail-order hatchery. What could be simpler than having a group of day-old chickens delivered right to your front door?

Purchasing through a mail-order hatchery can be particularly useful if you're interested in obtaining unusual or heritage breeds. Murray McMurray Hatchery, for example, sells rare breeds such as Phoenix, Buttercups, and Golden Laced Wyandottes, which can be difficult to find. One thing to consider when ordering through a mail-order hatchery is that the

There are many breeds of chickens from which to choose, and you'll want to consider all the characteristics and qualities that you're seeking before you making your choice. This is a Frizzle Cochin hen.

EGG COLOR

Just exactly where are those earlobes? They're located just below the circular area that is the chicken's ear.

While this may seem superficial in comparison to some of the other more serious criteria upon which you'll select your desired breeds of birds, egg color can be an important consideration to lots of people. The majority of chicken breeds produce eggs that are white or brown, but some breeds' eggs are blue, green, reddish, or even pink. So-called "Easter Egger" chickens produce a range of colored eggs, while Araucanas and Ameraucanas stay more restricted in the blue and green shades.

It is said that the color of a chicken's earlobes predicts the color of its eggs; though this doesn't always hold true, it's a guide. Where do you find those earlobes? On the side of the head, about where you'd think: look for either a white or red lobe, just below the ear. White lobes will produce white or lighter colored eggs and red lobes will typically produce brown eggs.

Colorful eggs boast the fun of novelty (what's more fun than collecting blue eggs from your very own hens?), but there's also a practical side: colored eggs tend to bring a higher price at farmers' markets than their white or brown counterparts. So if you're looking to bring in a little extra cash from your egg business, consider the appeal of colored eggs when selecting your breeds.

Eggs come in a variety of colors, including white and various shades of brown, blue, and green.

The White Leghorn rooster is what many people consider the quintessential American rooster.

minimum number of chicks you can purchase is often twenty-five, so if you don't have room for that many chicks (don't forget, they will soon grow into full-sized chickens!), you may want to choose a different means of purchasing chickens—one that will allow you to buy only as many as you need. Or you can partner up with a friend or two, each taking a share of the twenty-five chicks.

To this end, you might consider purchasing started pullets (young hens that are just about ready to start laying) either from a mail-order hatchery or from a local source. If you're only after a few healthy, mature, and well-bred chickens, this can be a particularly suitable option.

If you're unsure of which breeder or hatchery to buy from, consult other chicken enthusiasts in your area. Perhaps your next-door neighbor has had great success in raising chickens and can offer you advice on local sources for purchasing chickens. Or perhaps he has raised seventeen consecutive Best of Show chickens at the fair and would be willing to sell you a few examples of his best breeding stock. If not, local and national poultry organizations

The Barred Rock hen is a classic coop favorite.

may have advice on finding a suitable place to buy chicks. You can also consult your County Extension Service office; they will have good information and sources for you to consider.

In addition, the next time you attend a poultry show or a county fair, talk to some of the exhibitors and other knowledgeable people about sources for buying chicks or hens. Members of your local FFA or 4-H club may be interested in helping you find the perfect poultry. It is essential to make sure the chicks or chickens you're ordering are the healthiest ones you can find, so ask about health guarantees and make sure that you don't inherit someone else's problems. Sometimes getting free chickens from a friend isn't the best choice.

In essence, the best way to find a good place to buy chickens is to do some research and ask around. There are many, many excellent places to choose from, and as long as you have some idea of what you're looking for, finding suitable chickens should be no trouble at all. Good luck!

CHOOSING AND BUYING CHICKEN SUPPLIES

Even after you've planned your coop and selected your preferred breed of chickens, you'll still need to outfit your coop with the equipment and accessories needed to care for your chickens. Feeders, waterers, nest pads, lights, brooders, feed, containers in which to store feed—your shopping list can quickly grow quite long. You can purchase these items online

The Rhode Island Red rooster is a handsome fella.

The Naked Neck chicken is an odd-looking addition to your coop.

A Russian Orloff hen and chick scout for food.

Visit your local county fair or poultry show to locate breeders and to see what's available in your local area. This Silkie chicken is a fancy bird!

Finding the right chicken feeders, waterers, fencing, wire, and nest boxes is important. Whether you buy locally at the farm supply store or purchase through catalogs, ask around for advice from other people who raise chickens.

or through mail-order catalogs, or you can stock up at your local farm supply store. Bear in mind that while the selection from online retailers might be more extensive, shipping costs can add up. In this case, shopping locally may be more cost-effective.

When you're first starting out with chickens, it's easy to get caught up in the desire to do everything heartily and all at once—and that includes buying supplies and accessories. But until you've worked with and around your chickens for a while, it can be hard to know exactly which items will prove to be most useful to you. For this reason, you may want to start small. Buy only what you absolutely need to get started, and then work with your birds for a time. You may discover that the feeder style that you chose is less than efficient, or you may love the nest pad that you decided to try. Before investing significantly in vast quantities of supplies, experiment a bit to discover what you like and what works best for your situation. Again, asking other chicken experts in your area what equipment they like best may guide you to better decisions.

THE BEDDING DILEMMA

At this point, you're probably thinking, *What does choosing bedding materials have to do with building my chicken coop?* The answer is, nothing. Well, not exactly. But even if you build the nicest coop in the world, it won't be usable by chickens if it doesn't have a good layer of bedding to absorb moisture and provide a comfy floor for your birds—and a soft, clean place to lay their eggs in their nest boxes. So with this in mind, it's clear that choosing appropriate bedding material is not only an important step in preparing your coop for life with chickens, but a critical one as well.

There are many kinds of bedding to choose from, all with various advantages and disadvantages. Selecting the right one comes down to personal preference and budget as much as anything else, although some types of bedding are definitely superior to others in terms of easy cleaning, absorbing moisture, and such. Here are some of them:

- **WOOD SHAVINGS**: Widely available at feed and pet stores, wood shavings are easy to work with, excellent at absorbing moisture, and terrific at creating a soft bed for your chickens. They smell good and look good and can be a great addition to the compost pile after cleaning the coop. Depending on the number of chickens you'll be raising, they're fairly cost effective.

  In seeking the best possible wood shavings for your coop, there are a couple of considerations to keep in mind. In general, shavings made from softwood trees, such as pine trees, are the ideal, although some softwoods—such as cedar—are not recommended for chickens or other livestock. Although cedar shavings have a wonderful, woodsy smell that can help repel insects—an obvious benefit—there is a great debate among chicken keepers on whether cedar (and, to a much lesser extent, pine) shavings can be toxic to your chickens. Some people never have an issue; others have found the toxicity levels too high for young chicks. Unfortunately, little research has been done on the toxicity of cedar shavings when it comes to chickens, so if you intend to raise baby chicks, pine shavings—or another kind of bedding entirely—will prove more suitable.

  On a similar note, make sure that the wood shavings you purchase (regardless of the type) are free of any chemicals that could be harmful to your birds. Typically pine shavings sold for livestock will be safe to use.

Choosing the right bedding may take some trial and error on your part. You may try various products and materials only to end up choosing something straightforward, such as pine shavings. Pine shavings are clean to look at and smell nice as well. Your bedding will be used in your nest boxes as well as on the floor of your coop. The owner of this coop also adds in lavender, herbs, and edible flowers to her nest boxes for her hens to enjoy—and to help control mites.

One thing to avoid is sawdust, which—although similar to wood shavings—is composed of much finer particles (hence the name saw*dust*) that can cloud in the air and cause respiratory issues for your chickens. So breathe easy (both you and your chickens) and stay clear of sawdust in favor of wood shavings.

- **STRAW**: The main questions to consider with straw are what kind of straw to use and whether it should be chopped into pieces or left whole. In general, wheat straw is considered to be the best type of straw for chicken bedding, while oat and rye straw can be good choices as well.

  As far as chopping or not chopping go, there are advantages and disadvantages to both. Straw that has not been chopped can be difficult to work with, as it tends to tangle itself into large snarls that mix with chicken droppings and become a mess—yuck! By chopping your straw, you can avoid this issue, but chopped straw can be on the dusty side and could potentially lead to respiratory issues for your chickens. Straw doesn't have the absorbency of wood shavings which may also impact your decision.

- **DRIED LEAVES**: Dried leaves, gathered after they have fallen from trees in autumn, offer the advantages of being abundant and very inexpensive—in most cases, their only cost is the time you spend gathering them. However, dried leaves have the disadvantages of being prone to growing mildew and mold, and they tend to pack down and lose their fluffiness easily. Shredding the leaves can help prevent this. Pine needles can also make another good natural bedding choice, but unless you live in or near a pine forest, securing enough to use as chicken bedding could prove challenging.

  Always be sure that you choose leaves that are not toxic to chickens. The leaves of oak and buckeye trees, for example, contain toxins that could harm your chickens, so you will want to avoid them. If you're uncertain how to identify the poisonous leaves from the good ones, or are leery of making a mistake, a different choice of bedding material may be best for you.

- **SHREDDED NEWSPAPER**: If you are fond of reading newspapers, you know how quickly they can pile up after you've read them. Fortunately, shredded newspapers can make excellent bedding for chickens, so if you have a paper shredder, you can bid farewell to the ever-growing stack of newspapers and give them a new lease on life as chicken bedding.

  Shredded newspaper offers the advantages of being inexpensive (you're already paying for the newspapers, and you may be paying to dispose of them as well, so using them as bedding could actually save you money) and also somewhat absorbent, helping to keep your chicken coop nice and dry. In addition, shredded newspapers disintegrate rapidly and can be used as compost when their days as chicken bedding are over.

  There are a few disadvantages to shredded newspaper that should be considered. For one, if not changed often enough, it does likes to clump together, which can lead to a soggy and yucky pile of paper. This can become rather slippery, leading to undesirable side effects such as splayed legs in young birds. Fortunately, careful maintenance of the bedding conditions can help you avoid this issue.

  Another concern is the possibility of shredded newspapers containing toxic ink that could harm your chickens if they consumed it—and of course, chickens do have a tendency to peck at things, bedding included! If you're uncertain whether your newspapers feature nontoxic ink, a different choice of bedding may be necessary.

- **SAND**: Good-quality sand may not be the first thing that comes to mind when you think of chicken bedding, but it can be an excellent bedding choice for your coop.

However, unless you happen to possess a top-notch sand pit in your backyard (not likely), you will have to purchase builder's or construction-grade sand. The advantages of sand are numerous, and you may find that the ease and cleanliness of sand is worth the added cost.

The primary advantage of sand is that it is terrific at absorbing moisture and dries rapidly. Furthermore, sand is easy to work with and clean. What could be easier than scooping chicken droppings out of sand?

In essence, many different types of materials can be used as chicken bedding, and all can be perfectly useable under the right circumstances. Basically, as long as your bedding material is free of dust, mold, and moisture, you are well on your way to creating a soft, healthy haven for your poultry.

SELLING EGGS OR RAISING CHICKS

After your backyard or hobby farm chicken enterprise is in full swing, you might find yourself wondering what you can do to earn a small amount of income from your hens.

CHICKEN NEST PADS

No matter what bedding material you choose, it's bound to get somewhat messy, and if the thought of your hens laying eggs in old dried leaves makes you cringe, perhaps you should consider special bedding for your nest boxes. Many farming supply companies offer chicken nest pads, which are specially designed, one-piece bedding layers that are meant to be comfortable for your chickens while at the same time helping keep your eggs fresh and clean.

Chicken nest pads feature an array of tiny "tufts," coupled with holes in the center and in between each tuft, that discourage build-up of droppings while also keeping your eggs raised above the manure, reducing the number of contaminated eggs. Hooray!

Best of all, chicken nest pads are very easy to clean, thanks to their easily-removable nature. So if cleanliness is a major desire in your nest boxes, why not give chicken nest pads a try?

DEEP LITTER METHOD

No matter which kind of bedding material you eventually choose to use in your coop, when it comes to coop cleaning, consider trying the deep-litter method, which, when executed properly, can save you an incredible amount of time and effort, in addition to providing other valuable benefits for your chickens. Basically, the deep-litter method involves establishing a deep layer of bedding (several inches thick), and then adding to it—rather than subtracting from it—when cleaning the coop. For example, instead of scooping chicken droppings out of the bedding, rake them underneath the bedding material (where they can't cause you or your chickens any trouble) and add more bedding on top as necessary.

Over time, your layer of bedding will get deeper and deeper, with the chicken droppings and old bedding remaining at the bottom of the layer, where it naturally decomposes. Furthermore, the decomposition process releases a bit of heat, which is helpful for keeping your chickens warm during the winter months. Many people love this method for its time and convenience.

After a while—a few months, perhaps—you will need to give your coop a complete cleaning, removing all the bedding and starting fresh. But the time spent on this occasional chore is more than offset by the time you will save throughout the rest of the year. If you're interested in trying the deep-litter method, talk to other chicken enthusiasts that have experience with the system, and ask for advice and tips on how to create and maintain the bedding layer. If you do the job right, you will find the process to be time-saving, easy, clean—and maybe even fun. Okay, so fun might be stretching it a bit, but you get the idea!

SELLING EGGS

For the average chicken owner, the idea of home-grown, "fresh from the farm" eggs is a real appeal. From egg breakfasts and hard-boiled egg appetizers to cakes and other baked goods, there are plenty of reasons to love fresh, organic eggs. This brings up a question: wouldn't other folks love your eggs just as much? The answer, of course, is yes. Many people love the advantages of fresh eggs but don't have the property or time to pursue chicken keeping. This brings up a second question for the chicken owner: should you try selling your farm-fresh eggs?

No one can answer that except you, but there are some major factors and concerns to keep in mind that may influence your decision.

- **LEGALITIES**. Unsurprisingly, there are laws dictating the sale of farm eggs—although these laws tend to be somewhat less restrictive than those governing the sale of other farm products, such as milk. These laws vary depending on your exact location, so the best thing to do is to research your state and local laws regarding the sale of eggs before you get started. For instance, you may find that it's perfectly okay to sell eggs to your neighbors as long as they travel to your farm for the purchase, but you are not able to sell eggs in other locations without a business license. Also, whether you need a license in general to sell eggs may depend on the numbers you're producing; selling just a handful of eggs here and there may not put you in a category that requires licensing, but large sales might. Carton labeling is likewise variable; you may need to label your farm-grown eggs as "ungraded" and you may be able to reuse retail egg cartons—but you may not. And

Fresh eggs! You love them, and people in your area may be interested in purchasing the eggs your hens produce. Consider selling eggs if you're trying to earn a small amount of income from your chickens.

A bit of simple advertising may be all you need to find buyers for a small surplus of eggs.

Will you create a fun name for your egg "brand"? Keep in mind that any type of labeling (and what goes on the label!) may be subject to state or local laws.

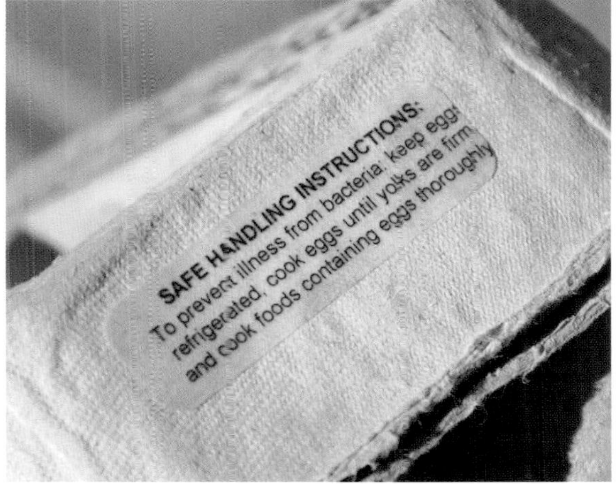

A sell-by date may be necessary (top), as well as other types of language (bottom), depending on the laws in your area.

if you decide to go into a larger-scale egg business, keep in mind that it will require a USDA inspection.

- **INCREASED FLOCK SIZE.** If your hens are proficient layers, they may already be supplying you with more eggs that you can use on your own. But even if this is the case, it's likely that you'll want to increase your flock size to make egg selling worth the effort. Are you ready for more birds? Is your infrastructure ready?
- **PRICING.** Naturally, it's unlikely that the small-scale egg farmer can compete with grocery store pricing and still turn a profit. But that's okay, because competing with the stores is not really your goal, or the goal of your local egg buyers. The goal is to deliver (or consume) a quality product from a local source, and buyers will likely expect to pay extra for this. Furthermore, some egg producers don't look at turning a profit per se; they figure that the sale of excess eggs is a way to pay for some specific part of chicken care—perhaps feed or bedding costs. You may enjoy the egg-selling process more if expectations are kept reasonable.

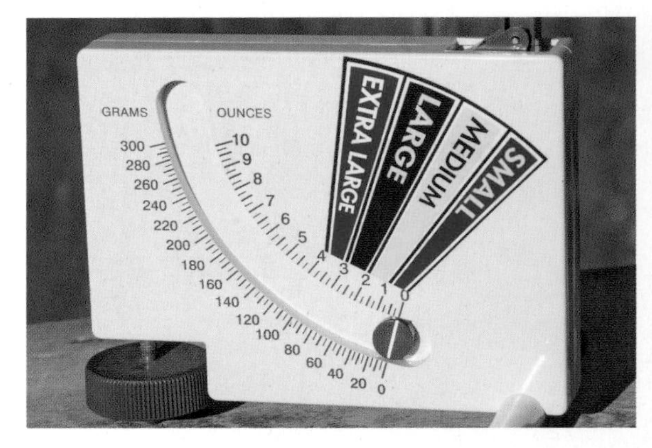

A simple egg scale is a useful tool for weighing your eggs and determining what size your hens are laying.

Are you prepared to handle the increased flock size that your new egg business may require? More chickens mean a larger time commitment.

Above left: Raising and selling chicks can be a fun and rewarding way for you to offset some of your chicken keeping costs, but it's not without effort, and you must plan ahead.

Above right: If you're going to raise chicks, you're going to need at least one rooster on your property. Embrace the farm-life experience and enjoy the crowing!

Left: You'll also need to talk with a poultry veterinarian about the possibility of vaccinating your chicks.

RAISING CHICKS

Eggs are one product that your chickens will produce for you, but don't forget the other: baby chicks! Many small-scale chicken farmers actually find a fair amount of local demand for chicks, and this can be a way to earn some additional income from your flock. Before you jump in, here are a couple of points to consider.

- **CHOOSE YOUR BREED CAREFULLY.** When choosing a chicken breed for your own use, you can consider only your own needs—eggs, meat, the physical size of

What chicken breed will you settle on for your chick-raising business? This White Crested Blue Polish hen (opposite) has a lot of fun personality in her appearance, as does this striking Spanish White Face rooster (left).

the birds, etc. But if the goal is to raise chicks that other people will want to buy, you should think about what chicken breeds are most marketable. As with other commodities, chicken breeds go in and out of style, and to have success selling your birds you'll want to have a desirable breed that other people will want to add to their flocks. Similar to the way that many hobby-farm gardeners grow rare heirloom vegetable varieties, many small-scale chicken breeders are opting to raise unique heritage breeds. Doing this can help you stand out from the crowd, and it provides the added benefit of helping to maintain breeds that suffer from dwindling populations. That said, some potential buyers will be more interested in acquiring one of the more popular chicken breeds, so you may want to stick with the major, well-known breeds unless you have a specific interest in heritage varieties.

- **KEEP DIFFERENT BREEDS SEPARATED.** Prospective chick buyers may want purebred chickens. Raising purebreds is simple if you keep only one breed, but if you have multiple breeds of chickens and roosters, steps must be taken to separate the breeds (or, at the very least, keep them separated during the times when you are actually actively raising chicks!). This can be challenging and add labor to your workload, so you may want to select only one breed and stick with it.
- **AVOID SELLING ONLY PULLETS.** Some buyers who are looking to get into raising chickens themselves may ask you if you will sell just the female chicks. While this might make for some happier customers, taking this route will eventually cause you to be overrun with roosters. It may be worth your while to simply sell groups of random chicks to prevent this from happening.
- **BROODY HENS ARE A PLUS!** Brooding is simply a behavior that hens demonstrate when they "feel" as though they should be tending to a clutch of eggs. When a hen is brooding, she stops producing eggs, sits on the nest box for long periods of time without leaving, and often shows changes in her personality, becoming pushy toward the other hens and possibly more irritable toward people. Hens with strong brooding instincts may brood and attempt to sit on their (or another hen's!) eggs for up to 21 days—the normal chicken egg

Your flock of chickens may enjoy hanging out together, but if you're trying to raise a specific breed, you'll need to separate chickens into groups.

A beautiful trio: a Cochin Frizzle rooster, hen, and chick. Cochins are a breed that is known for being broody, which is a benefit if you want to raise chicks.

If you choose to raise chicks, you'll also get to observe and enjoy moments like these—a hen carefully guarding her brood outdoors.

incubation period—regardless of whether the eggs are fertilized. A strong brooding instinct is exactly what you want if you're trying to raise chicks, but not what you're looking for if you're the average chicken owner who just wants a regular source of eggs, since egg production drops with broody hens (plus they get cranky!). Whether a hen shows broody tendencies depends on the particular bird and her breed. Some chicken breeds have been bred with an eye toward egg production, and as a result, some of these hens may have lost some of their brooding instinct. A few breeds that are still excellent "brooders" include Silkie, Cochin, Bantam, Old English Game, Frizzle, and Sussex, for starters.

3 DESIGNING YOUR COOP

If you're interested in the nitty-gritty process of designing a coop to your own plans and specifications, then this chapter is for you. We will guide you through the major areas you need to consider when designing your coop, including size, ventilation, and roof types, as well as many other details that are important to understand. Upon reaching the end of this chapter, you'll have a better idea of what your coop will need to be the perfect abode for your flock.

THE COMPLEXITY OF YOUR PROJECT

Although it might seem great to build a gigantic coop with every bell and whistle you can think of, before proceeding any further, it would be wise to sit down and ask yourself the following question: *How complex a project can I reasonably undertake?*

After all, if a bad gardener has a black thumb, it could be said that a poor carpenter may soon have a black-and-blue thumb! So if

Above: There are many decisions to make when designing your coop. How much room will you need for your birds? In this case, the hens have plenty of room in their coop and run; they're also protected by the appropriate hardware cloth wire.

Opposite: Designing the perfect home for your chickens can be a lot of fun. Your chickens will benefit from the ability to keep safe from predators as well as having protection from inclement weather.

your carpentry skills are limited or nonexistent, and you will be working on your coop for only a few hours a day on weekends, it would be wise to scale down the chicken coop of your dreams into a project that is more easily achievable. Instead of ten nest boxes, maybe three is a more realistic goal. This is a perfectly fine decision, and with this in mind, the chicken coop instructions offered in Chapter 5 are geared toward someone with basic carpentry skills and tool knowledge.

On the other hand, if you (or a family member or friend) have some experience in working with lumber or power tools, and have the time to commit to building an extra-special coop, then feel free to consider all the aspects of chicken coop design and customize your coop so that it is exactly the way you want it.

THE SIZE OF YOUR COOP

What size will you make your coop? This is an important decision as it will determine the number of chickens that you'll be able to comfortably house. Do you need a large coop? Will a small coop suffice? Something in between?

Generally speaking (and remember, the more space per bird, the better), you'll want a minimum of 3 to 4 square feet of interior space for each hen in your coop. This means that a coop with 12 square feet of interior floor space will have enough room for three to four birds.

Now, three to four birds is by no means a sizeable flock, but for the novice chicken enthusiast, it can be the perfect number with which to get started. But what if you already have extensive experience with chickens

In this case, the hens are vulnerable to predators due to the use of the flimsier chicken wire on their run. They're also overcrowded with too many birds in too little space. This can cause health problems as well as problems with birds pecking each other.

Chickens love to roost and even free-range chickens will find places to perch. This flock is passing the time in lilacs.

and are ready to establish a more populated flock? A larger coop with more extensive areas of enclosed space will provide room for an increased number of birds. Simply calculate the projected square footage of the enclosed area of the coop and calculate the number of potential birds that could inhabit the coop.

Remember, you don't want to overcrowd your coop in an attempt to squeeze in more birds; insufficient elbowroom can cause serious problems for your flock, beginning with pecking at each other and fighting, and possibly escalating all the way to cannibalism. Avoid these potential problems by providing plenty of space for your chickens and don't overcrowd them.

You should also consider the future size of your flock. Maybe you only have four hens right now, but how about next year? Do you foresee yourself significantly expanding your flock, or are you happy with a smaller group of birds? If you anticipate an uptick in your chicken population, then it might be wise to accommodate for that possibility when you sketch your design. You don't want to go to all the trouble of constructing a custom coop only to decide that four chickens really isn't enough and eight would be so much better. Do enough research ahead of time so that you can build the coop you want in the size you need.

ROOSTS

Like most birds, chickens prefer sleeping off the ground, either in trees, on top of their coop, or on roosts. Because sleeping outdoors makes your chickens far too vulnerable to predators, you undoubtedly want to keep them in the coop at night, and that's where roosts come in.

Having enough roost space is important. Each chicken needs between 8 and 12 inches of roost space, and staggering the roosts (as shown here) gives each bird plenty of comfortable roost space.

Wooden roosts, whether poles or boards, are most desirable. This is especially true if the wood isn't sanded to perfection and has a little roughness to it, as that makes it easier for the chickens to grip. Rounded corners can be helpful, but it's also very important that the roosts be thick enough for the birds to perch comfortably—a minimum of 2 inches. Never use metal roosts; chickens have trouble gripping metal, and if it becomes too cold, your chickens could get frostbite.

In addition to making sure that your roosts are of the appropriate size to suit your chickens' feet, you must also provide enough roost space to accommodate all your chickens; each bird will need a minimum of 8 inches of length to avoid crowding—and more is better. Make 1 foot per bird the minimum and your chickens will thank you.

Another thing to consider is the height of your roosts in relation to the nest boxes. Roosts should be about 1½ to 2 feet off the floor, and they should not be placed above your nest boxes. Otherwise, when the hens roost there, they will add a mess just where you don't want one. The roosts can also be staggered in grandstand fashion for cleanliness.

NEST BOXES

It would seem logical to think that each hen needs a nest box all to herself, but the truth is that one box will do nicely for three to four hens. Some even say that it doesn't matter how many nest boxes you have; all the hens will want to use the same nest box.

But for the purposes of designing your coop, the number, size, and location of your nest boxes does matter—a lot. Many of the concerns about nest boxes have to do with keeping the eggs clean—in fact, that's the main reason you will want nest boxes. Nobody wants to eat a dirty egg, much less a cracked one, and that's exactly what you may get if your hens lay their eggs on the floor or anywhere but in a designated, clean nest box.

As for size, ideally your boxes should have 1 square foot of floor space (12 × 12 inches), and be 12 to 14 inches high or thereabouts. The roof or ceiling of your boxes needs to be slanted, like a shed roof, to discourage birds from roosting on top of them and (yes) making a mess.

If your nest boxes are bumped out on the exterior of your coop, there will be more room inside the coop for all your other chicken paraphernalia, but that kind of box will need a very good roof to prevent rainwater from leaking into the nests.

There are many different ways to go about providing nest boxes for your hens. You'll need to make sure your nest boxes are big enough and that you provide an adequate number for your flock. These nest boxes are being well used.

BROODERS

There are plenty of confusing terms related to chickens and chicken keeping, and you've stumbled onto one of them! A **brooder** is specifically for the raising of young chicks you've purchased. These chicks won't have a hen to keep them warm, and they can't be allowed to be in the same coop as your older chickens until they grow big enough to defend themselves; that is where the brooder comes in. A **brooding area** in your coop is for chicks that are still with their mother and is usually just a separate portion of the coop. Another confusing thing is that the heat source inside your brooder may also be called a brooder in some cases, be it a heat lamp or some other type of heating system. In this book, whenever you see the word *brooder*, we will be referring to the place where you are going to raise your motherless chicks.

The brooder doesn't have to be elaborate, but it does need to be warm and clean. Your chicks are going to need lots of TLC during those first days at home. A spare room in your barn or home can be an ideal place for your brooder. This tack room works perfectly; it keeps out predators and allows the chicks to stay toasty warm.

You must order a minimum of twenty-five chicks from most mail-order hatcheries, so if you don't need that many, try teaming up with a friend or two to share the order.

Brooders have been constructed out of pretty much everything under the sun: cardboard boxes, rubber tubs, plastic children's wading pools, and of course wood, although a cardboard or wooden box is most often recommended. One kind of brooder is called a **battery** and is actually several tiers of brooders stacked up, each with its own heat source, but unless you intend to raise dozens of chicks, you probably won't need anything so elaborate. A single box will do nicely for one batch of chicks.

Now that your brain is reeling from all this information and you're vulnerable to a surprise attack, we're going to throw another mind-boggling fact of chicken life at you: your brooder is probably not going to be in your coop.

"What?" you exclaim. "This is a book about building chicken coops, and brooders don't usually even go in chicken coops? What gives?"

These rare Russian Orloff chicks are thriving in their brooder. Make sure you keep your brooder at a temperature of at least 90 to 95 degrees to ensure the health and well-being of your tiny charges.

But in a way, you could consider a brooder as a mini-coop. You need to provide your brooder with many of the same things as your coop: feeders, waterers, adequate bedding and ventilation, and predator protection. But you must also provide special amenities for the chicks—the most important being a heat source.

Without a heat source, your chicks couldn't survive. Warmth is vital to their well-being, and they must have 24/7 access to either an infrared heat lamp or a special heater specifically for young poultry. Most heat sources are suspended above the brooder; others go inside it.

Chick feeders are a convenient way to provide food in your brooders, but you have to be careful how you supply your youngsters with water. Wet litter can bring on coccidiosis, which is potentially fatal to young chicks, so if you use a traditional waterer, be sure to change out any wet litter as quickly as you find it. Another option is to use a waterer fitted with poultry nipples—kind of like the water bottles used in rabbit hutches.

Then there's the great bedding debate. What should you use for litter in your brooder? Some of the options are the same as for your regular chicken coop, such as wood shavings, sand, or chopped straw, but there are a few kinds of litter you need to avoid in your brooder. For instance, shredded paper is not a good choice for chicks, because they have trouble maneuvering through it; the same goes for whole straw. At the beginning, cover the litter with paper towels so your chicks don't peck at it, but after a couple of days, once the chicks are eating well and using their feeders, you can take the towels out.

As for the location of your brooder, many people keep their brooders inside their homes until the chicks reach a certain too-big-to-be-in-the-house age, and you may or may not want to follow suit. For one thing, chicks kick up a lot of dust. For another, unless their brooder has a roof or high-enough walls, they can escape. But being in the house certainly does cut down on the threat of predators, barring overly curious house pets. Oftentimes a warm utility room or a small spare room can work well. Many people use an extra room in their barn or in a small heated shed.

If you keep your brooder inside your home and you have pets such as dogs or cats, cover your brooder with a roof made of hardware cloth. This will provide sufficient ventilation and predator protection at the same time (yes, Poochy and Meowzers can be predators). For some fun and to learn what not to do, watch the old *I Love Lucy* episode entitled "Lucy Raises Chickens."

STATIONARY COOPS VERSUS PORTABLE COOPS

As you've probably gathered, designing a chicken coop requires a lot of decision making. Like most aspects of farming, there isn't just one cut-and-dried way to do it. Each aspect of your chicken coop requires careful consideration and evaluation. That's why we're presenting you with another decision to ponder: stationary coops versus portable coops. What's best for your setup?

STATIONARY COOPS

The best attribute of a stationary coop is that, in all likelihood, it will live up to its name and you will never have to move it. So feel free to make it big, roomy, and robust. You can use heavier, more durable materials and install permanent predator protection, such as laying brick pavers around the perimeter of your run or burying wire. To top it off, you will be able to keep more chickens because a bigger coop naturally means more space for more chickens!

The best attribute of a stationary coop, however, can also be its worst attribute, as you won't be able to change your mind about its location in two months or six years, unless you want to go to extensive effort to move the coop. This can be a serious drawback for some people.

However, if you're of the mindset to let your chickens free range during the day, then a permanent location for your coop may make a lot of sense. You can add landscaping around it and beautify it so that it adds to the attractiveness of your yard or property.

PORTABLE COOPS

The word *portable* can be a little misleading when it comes to chicken coops because the actual degree of portability varies depending on how you design your coop. You might want it to be hand-portable—easily moved by a few

A stationary (or permanent) coop is one that is built and stays in one location rather than being easily moved around. Larger coops are typically permanent, and they have the ability to be wired for electricity—a nice convenience.

This large coop is an exception as it was placed on a wagon bed and is easily moved around via the farm tractor. This is a nice way to take advantage of new grass every few days, but a great deal of room is required to maneuver this type of coop

This coop is sized for about fifteen hens, but it is still portable due to its wheels. The owners move this coop to a new location once a week.

Smaller coop kits have the advantage of being lightweight and easily moved around by hand. Their main disadvantage is that they are not as predator-proof as a permanent coop.

people. But a portable coop can also be one that is built on wheels or on the bed of a wagon, a tractor, a truck, or (if it's small enough) one an ATV can pull.

In any case, a portable coop must be smaller and more lightweight than a stationary coop. You may not be able to have enough space for many chickens in a portable coop as in a stationary one. An even greater concern, perhaps, is that you might not be able to provide a level of predator protection that is as high as that of a permanent, stationary coop. This is a more serious disadvantage than most.

The best aspect of a portable coop is that you will be able to move the coop whenever you wish, allowing your chickens to enjoy new grass and places to forage.

DESIGNING YOUR COOP WITH EFFICIENCY IN MIND

Now it's time to get down to brass tacks in your coop design. Good maintenance of your chickens and coop is essential, and it's nice to make your day-to-day care to be as efficient as possible.

Here's an irrefutable fact about poultry: they're messy. Bedding, feathers, sand,

The large cleanout door on this coop makes accessing its interior easy, which makes cleanup less of a chore.

CHICKEN TRACTORS

Are you looking for an extremely easy setup for your chickens? Do you want a safe place for your chickens that doesn't require much actual construction? If so, and if your flock of chickens is relatively small, then a chicken tractor may be just the kind of chicken coop you've been looking for.

A chicken tractor is a portable chicken coop in which the actual coop and the run are securely connected together in an all-in-one unit. To facilitate frequent moving, chicken tractors have wheels or handles that allow you to move the coop around your yard to give your chickens continual access to fresh grass. Because chicken tractors must be easily moveable, they are somewhat limited in terms of size, meaning that they are not suited to someone who needs a coop for a large flock of birds.

At the same time, a chicken tractor must include the same features of its more traditional relatives, including a secure roof and durable walls to keep your chickens safe from predators. However, because chicken tractors are not tethered to the ground—and are frequently moved around your yard—they cannot offer the same cold-weather protection as a stationary coop, so if you live in a region where the temperature frequently drops below freezing, a chicken tractor might not be suitable for your circumstances.

In addition to being an excellent option for small-scale chicken raising, chicken tractors can also be used for quarantining new or ill birds in your flock (more about that on page 68).

Chicken tractors—due in part to their small size—may not require as much construction as stationary coops and are relatively simple to build. In some cases, you might even be able to repurpose a small structure, such as a doghouse or large rabbit hutch, into a top-notch chicken tractor. In short, a chicken tractor can be a great solution for those who can't commit the necessary time to building a large coop but would still like a versatile and easy-to-maintain shelter for a small flock of chickens.

© Cathleen Abers-Kimball / Shutterstock

droppings, spilled food and water—your pristine new coop won't be pristine for long and it must be cleaned regularly. But when you keep efficiency in mind during the design stages, it doesn't have to be difficult to restore cleanliness to your hens' abode.

For the most efficient setup, you'll want a special cleaning door on the side of your coop that can be opened when you need to clean up droppings or change the bedding. As you'll see in Chapter 5, our project coop has not one but two ways to access the coop for cleaning: there's a small door on the side for quick tidying jobs, and the entire front of the coop is easily removable for major cleanings. Removing the front of the coop isn't something you're going to do regularly—most jobs can be accomplished via the small door on the outside—but when you really want to get in and do a thorough job, you'll appreciate the removable front wall. Some coops are designed with large doors; basically an entire side of the coop is on hinges and opens for thorough cleaning. This is super handy.

As for the nest boxes, you can clean them the same way you collect eggs from them, through the special door that opens to the outside of the coop. For extra simplicity, angle the floor of your nest boxes with a slight slant so the nesting material will slide out more easily when cleaning.

One option to consider during your design process is adding a linoleum floor for your coop. This can significantly increase your coop's ease of cleaning. Your choice of bedding goes on top so the slippery surface isn't an issue. You just lay it loose in your coop and it lifts out to make cleaning a breeze.

Being able to access your nest boxes and collect the eggs without having to enter your coop makes it super easy and quick to bring in each day's eggs.

EFFICIENT EGG COLLECTING

Eggs are one of the biggest perks of chicken keeping. Hatching eggs and watching as the adorable chicks emerge is an amazing experience, of course, but it's a wonderful thing to be endlessly supplied with a multitude of eggs for culinary purposes too. The satisfying feeling of self-sustainable living is felt most when producing food, especially eggs from your very own chickens.

But those good feelings can be dampened if gathering the eggs is frustrating for you. If your chickens are laying anywhere they want or if they're leaving you broken or messy eggs, you'll want to think about eliminating as many of these issues as you can. Make it easy to collect eggs by incorporating an outside access door to your nest boxes. A simple flip-down door with a latch, located on the outside of your coop, will make it possible to gather easily eggs without having to enter the coop, which makes this job quicker, easier, and cleaner for you.

Having the forethought to plan for egg-collecting days will come in handy when mapping out your chicken coop's floor plan. You'll be glad you gave the issue the attention it deserves when you're ready to collect your first eggs.

Using a special egg basket makes gathering your eggs a little more fun. This English wire egg basket adds a lovely accent to egg collecting.

It's helpful to have the ability to separate birds for various reasons. If one bird is being pecked on by the others, it can be helpful to isolate her for a time until they're used to having her around.

This owner is using a smaller extra coop to separate birds of different ages. The older birds use the outdoor runs and the larger coop.

INCORPORATING SEPARATION AND QUARANTINE AREAS INTO YOUR COOP

As happy as your flock probably is, there will be the occasional time when one or more of your chickens will need to be separated and/or isolated from the rest of the flock. Whether this occurs as a result of quarantine, sick birds, or simply a broody hen, you'll want to make things easy on yourself by incorporating a quarantine area into your coop design.

One way to prepare is to construct a divided coop. It is just as the name implies—a coop divided into two or three sections for the purpose of separating groups of chickens, including for health-related issues. The walls can be made out of various materials, usually types of wood and/or hardware cloth, or even chicken wire. Some divided coops have walls with doors in them, but it's up to you. Doors inside the coop are more convenient, but it can be safer to simply have two outside doors leading to each section of your coop.

Another option is to build a secondary coop. This can be identical to your primary coop, or it can be something smaller and simpler, such as a chicken tractor. The latter option is particularly sensible, for even if you possess a large flock of chickens, it is highly unlikely that you will need to separate more than a handful of them at any given time.

So why might you need to separate your chickens from each other? Well, many circumstances could lead to the rearrangement of your flock. For example, a broody hen (see next page) needs a private home while she incubates her eggs. This location needs to be safe from predators and from other hens who may be interested in claiming the nest for their own. A simple brooding pen will suffice for your broody hen—a miniature coop just for the hen and her soon-to-be chicks.

What if one or more of your chickens should become ill? Sick chickens need to be kept away from their healthy compatriots in order to prevent the further spread of the disease. Utilizing another coop—or using isolated space in your divided coop—is a helpful solution.

Any time you purchase new chickens, it's wise to quarantine them for a time, just in case they are carrying any diseases that could be spread to the rest of your flock. Two to four weeks is the standard length of time that new chickens should be under quarantine. Chickens should also be quarantined for a few days after attending a poultry show in case they managed to pick up an undesirable ailment while out and about.

SEPARATING BROODY HENS

A broody hen is a marvelous example of maternal magnificence. Just think of all those innate instinctive qualities that suddenly manifest themselves when a hen decides to raise a clutch of eggs—it's amazing!

But exactly what happens when a hen "goes broody"? And what does it mean for your chicken coop project? Well, as your hens go about their daily routine, they lay eggs that you collect. When a hen goes broody, she wants to sit on a nest of eggs. Whether she carries this project out long enough to raise chicks will vary from one individual to the next, but regardless of the end result, a broody hen becomes somewhat territorial and obsessively focused on her nest.

If a hen does stick with her nest and raise a clutch of eggs to chickhood, she will be essentially confined to her nest for a three-week period. Aside from a daily jaunt to take care of important activities such as eating and drinking, she'll remain on her nest nearly without ceasing.

In this case, the separation is being done inside the coop run underneath the raised main coop. Here is where chicken wire comes in handy; it effectively provides an easy place to keep a hen and her chicks until they're old enough to join the main flock.

Hens that are sitting on a clutch of eggs are said to be brooding. Sometimes you can use a crate to house your broody hen. This provides her with everything she needs while giving her the ability to remain isolated from the other hens.

For this reason, it's easy to see why you wouldn't want her to establish her brooding project in one of the nest boxes in your coop. In a busy coop with multiple hens utilizing the same nest boxes on a daily basis, a broody hen just doesn't have the privacy that she wants and needs in order to properly raise her eggs; nor will you want one of your nest boxes taken up by a long-term occupant. Additionally, you'll want a safe and isolated location for the hen and chicks, so plan to accommodate broody hens separately.

You'll soon discover that only some hens go broody—others never seem to think about it. But you'll likely notice the same hens repeatedly going broody over time.

The safety of your hen and her eggs (and later on, chicks) is paramount, so make it a priority to ensure that the accommodations for your broody hen are 100 percent predator-proof. Your hen will be sitting—vulnerably—on her nest for a minimum of 21 days. Keep her stress levels low by protecting her from predators and providing a safe, enclosed area that is outfitted with a nest box, an area for her to use when relieving herself, and an area with food and water.

The location and appearance of your brooding area can vary according to your situation and specific needs. You could convert an existing structure into a brooding area—perhaps a fully enclosed, empty box stall in your barn could be converted. Or simply amend your coop design to include a separate brooding area that will allow the broody hen to stay in proximity to her friends while still maintaining her privacy and need for peace and quiet. (More on this in Chapter 6.) Large dog crates are sometimes used as brooding areas; hens hardly move from their nest when broody, so expansive amounts of space are usually unnecessary until the eggs hatch, and you can place the crate inside your run, if desired. This allows your broody hen to benefit from the protection of the coop while remaining isolated from the other hens.

One important key: keep the brooding area very close to or on the ground. This is for the safety of the chicks after they hatch. When they begin toddling out of the nest, you don't want them to fall any distance to the ground.

RUNS

Unless you choose to free range your chickens (and maybe even if you do), you will want your coop to have a run. A **RUN** is a fenced-in area attached to the coop that gives your hens safe and easy access to the outdoors, allowing them to go in and out as they please. Of course, free-range chickens don't necessarily need a run on a day-to-day basis, but it's a good idea to have one available if you can't be around to supervise your chickens during the day.

One of the main things to consider when designing your run is—as you have probably guessed—size. It is important to provide enough space for your chickens to prevent them from getting testy with each other. Overcrowded chickens are unhappy chickens, so it's best to avoid the situation from the get-go by making your chicken run large enough to provide approximately 10 square feet of space per bird. The height of your fence depends on whether you want to put a roof on it. If you go without a roof, your fence needs to be at least 6 or 8 feet tall to prevent predators from jumping inside—but this leaves your chickens exposed to hawks and other birds of prey, who aren't deterred by *any* height of fence.

Most runs are constructed from some type of wire fencing, but don't go out and buy yourself a roll of chicken wire just yet. If somebody held

The broody hen can be housed in her coop, but she will benefit from the extra safety and seclusion of her crate.

DUST BATHING

Just as pigs need to wallow, chickens need to dust-bathe. It's beneficial to both their health and their happiness, helping to keep them cool and free of parasites. It's also part of how chickens socialize with each other, sort of a poultry beach party. If your chickens are allowed to free range, they will probably shoulder the responsibility of finding dry, sandy areas to roll in themselves, but you can certainly provide a designated dust box or an area for them if you so wish or if your coop/run situation makes it necessary. This can be as simple as digging up a bit of grass in the yard or providing a box filled with sand or soil. To be perfectly honest, good old sand can't be beat for your dust-bathing area. Chickens enjoy dust bathing and it's imperative to supply an area for this activity to promote the health of your chickens. Your hens will thank you!

Right: You can provide a sandbox full of construction-grade sand for your birds. They'll thank you for it and enjoy it every day.

a contest to find the type of fencing with the worst misnomer, chicken wire would take the grand prize. Yes, it's been used for chicken runs forever—but if you'd like to have chickens for any length of time and keep them safe from predators, you'll want to find a better option.

The truth is chicken wire is just not going to serve as a strong enough barrier between your chickens and a predator's claws or talons. Chicken wire is extremely flexible because of the large gauge of the mesh. Smaller predators, such as weasels, can simply crawl through to snatch your chickens, while larger ones, such as dogs, could even tear it apart.

The best choice for your run is hardware cloth. Hardware cloth has a smaller gauge of mesh, so it's stronger than chicken wire. Galvanized ½- or ¼-inch hardware cloth is recommended for chicken coops, but ½-inch is more generally used, as it is somewhat stronger. When it comes to protecting your hens, you want the strongest fence available to you.

LIGHT

Lighting is another aspect to consider when planning your coop. Windows are an obvious means of letting light into your coop, and certainly the easiest as well. Natural sunlight is not only free (saving you the cost of electricity), it also serves as a disease killer, helping to sanitize the interior of your coop and keep your chickens healthy. In addition, windows are very helpful for ventilation, as we will discuss in further detail in the next section.

However, natural lighting has two disadvantages. The first is that you can't control it. When the sun goes down for the night, so too does your ability to work in your coop. This can be a major problem if you live in an area with relatively few daylight hours during the

These portable coops have an electrified run that does a great job of keeping the chickens in and predators out.

CHICKENS AND THE GARDEN

If you enjoy gardening, you might wish to consider giving your chickens access to your garden. Your chickens will enjoy the opportunity to browse your plants and you'll gain the benefits of natural pest control. This can be a real side advantage of keeping chickens on your property and in your garden. Chickens love to peck and eat insects, and you'll find that chickens are not only a great defense against garden pests like beetles and slugs, but that they also take care of pests like ticks that can be a concern for you and your other pets. Besides ridding your garden of pests, you may find that the chickens provide valuable assistance with weeding and, to a lesser extent, mowing!

So, if this sounds like a good option for you and your feathered farm hands, make sure your garden has a selection of chicken-friendly plants. Some of the most popular choices for chickens are nasturtium, broccoli, sunflowers, many kinds of berries, and leafy vegetables such as kale or cabbage. If you'd prefer not to have chickens in your garden but still want to grow food for your chickens to enjoy, consider container gardening some special plants just for them!

Growing food in your garden for your chickens to enjoy means double the fun. Gardening is a wonderful pursuit, and what's better than combining it with your love for chickens?

There are a number of benefits in allowing your chickens to have access to your garden, including control of pests and possibly weeds.

If your coop is large enough for a window, it's a wonderful addition for your birds. It allows natural daylight as well as warmth into your coop.

winter. For example, if the sun sets at 5 p.m. but you don't get home to tend to your chickens until 6:30, you're going to have to do so by way of a flashlight, which certainly makes things more complicated.

The other disadvantage to natural lighting—and it's related to the first—is that, during the winter, the light can be insufficient for promoting year-round egg laying by your hens. Naturally, chickens lay eggs during the warmer months when the days are long and cease laying eggs as winter approaches and the days get shorter. Think about it: If you were a chicken, would you want to be raising a flock of chicks in the winter? But if a steady stream of eggs for the breakfast table is something you desire, then you'll want to encourage your chickens to lay eggs during the winter. Fortunately for you, chickens can easily be convinced to continue laying eggs even during winter. You accomplish this by lengthening their "days" using electric lighting.

Adequate lighting is a real plus in the coop, especially if you wish to maintain your hens' ability to lay eggs during the winter. They'll need around fifteen to sixteen hours of light per day during the winter months, so an additional light source will obviously be necessary.

By installing electric lighting in your coop and using a timer, you can turn on the lights during the night to effectively increase the length of each day, causing your chickens to continue laying eggs even as winter sets in. To make certain your chickens continue laying eggs year-round, make sure your coop is lit—either by the sun alone or by a combination of the sun and your electric lights—for at least fifteen, maybe even sixteen, hours per day. Be consistent and provide lighting each and every day.

Although it seems like a good idea to have your lights turn on in the evening, it may prove better for your chickens if you set the timer to turn the lights on in the morning, before the sun has risen. This way, you allow your chickens to settle in on their roosts before darkness arrives. Replacing this gradual sunset with electric lighting—which goes from light to dark at the flick of a switch—would leave your chickens in the dark and unable to see where they are before they are settled down for the night, so avoid this if possible. By turning the lights on in the morning, while they are still settled in sleep, you shouldn't have any issues.

An exterior electric light on your coop is beneficial when doing chores in the early morning or after dark.

And of course, in addition to encouraging your hens to lay eggs all year-round, artificial lighting will enable you to tend to your chickens after dark if needed—hooray!

VENTILATION

Ventilation is a highly important aspect of coop design. Without proper and appropriate ventilation, your flock is at increased risk of developing respiratory disease. So for the health of your birds, you'll want to be sure that your coop has ample ventilation.

But how do you go about that? The amount of ventilation is directly correlated to the size of your coop; some experts feel that ventilation should comprise a minimum of one-fifth of the wall space in your coop area. Others cite figures of 1 square foot of ventilation per bird.

In establishing good ventilation for your coop, one of your main goals should be to try to prevent the buildup of gases such as ammonia and carbon dioxide. Ammonia is a foul-smelling alkaline gas that is released by chicken droppings. In small amounts, ammonia will not pose any problems to you or your chickens, but in large amounts, ammonia can be toxic and could even kill your chickens.

Fortunately for you, ammonia—like the friendly helium gas used in birthday balloons—is lighter than air, which means that it will naturally drift upward. As a result, if you build your coop with a large ridge vent or another type of opening in the ceiling, you are providing an exit through which ammonia gas can float up and drift harmlessly into the atmosphere. Yay!

Carbon dioxide is the gas that is released during the breathing process by humans, chickens, and many, many other animals. It is completely natural for it to be present in the atmosphere in small amounts, and it is not harmful in and of itself. However, if your coop does not have adequate ventilation, carbon dioxide levels will increase. As a result, carbon

Using hardware cloth on the gable ends of your coop adds lots of great cross ventilation during the summer. Wooden inserts can be made to cover the ends when colder weather demands it.

The ability to open a window in your coop on hot days will aid in keeping your flock cool. Always add hardware cloth to the inside of the window to prevent access from other animals or birds.

dioxide will slowly crowd out the oxygen, potentially reaching dangerous levels.

Furthermore, the air in your coop can grow stale if it isn't frequently refreshed, and stale air is detrimental to the health of your chickens. So as you can see, proper coop ventilation can go a long way toward maintaining the well-being of your flock.

There are many options available for creating and optimizing ventilation in your coop. One obvious way to establish ventilation is through the use of windows. What's better than a nice summer breeze wafting through your coop, eliminating undesirable gases and bringing a nice supply of fresh air to your poultry?

Adding a simple vent to your coop allows additional ventilation. Vents are easy to install and effectively keep out unwelcome guests.

But there are several things to consider when designing windows for your coop. Ideally, you should have windows that can be opened and closed, so you can control ventilation on cold days when you want to keep your coop warm. Also, you'll want to have your windows covered with hardware cloth, so predators cannot gain access to your chickens.

Another good way to ventilate your coop is through the use of vents or openings—screened or protected in some other fashion, of course—located close to the ceiling, preferably on the north and south ends of your coop. These vents, thanks to their positioning at the top of the coop, will allow the release of warm air, which, like ammonia gas, naturally rises upward. It is an interesting fact of science that warm air can hold more moisture than cold air, so by providing an opening through which warm air can exit the coop, you are also providing an excellent means for the removal of moisture from the coop, which can minimize the spread of mold and disease.

You can also remove moist air from your coop by establishing cross ventilation—in other words, opening the vents or windows on opposite sides of the coop so that air can flow in the coop at one end and exit at the other. You can always add fans to your coop and use them on those hot days of summer when your chickens need extra ventilation.

In addition to windows and vents, it might be wise to have at least one of the walls in your coop feature a large area of hardware cloth for maximum ventilation. In other words, rather than building a wall out of wood and leaving a window opening, you might forego the wall entirely and just cover that side of the coop with hardware cloth. After all, what could provide better ventilation than an open wall?

If you live in a region with cold winters, you can seasonally cover the areas of hardware

You can winterize your coop by adding wooden inserts to your run. They provide wind and rain protection to your hens and keep the coop and run a bit warmer during those frigid days of winter.

cloth with an insulating material such as wood or metal. As beneficial as ventilation can be for your birds, drafts of cold air during the winter months are *not* desirable. Thanks to all their feathers, chickens can tolerate rather cold temperatures fairly well, but significantly cold temperatures (compounded by cold drafts) can cause frostbite or frozen combs. Of course, your chickens can suffer these ailments even without drafts if the overall temperature of your coop is too low, but drafts will definitely increase the chances of your chickens developing these issues.

There are many ways to protect your chickens from drafts, and you can use some or all depending on your circumstances and coop design. The simplest way is to close the windows or vents that are letting in drafts, locking the cold wind outside and relying on the remainder of your ventilation openings to provide fresh

Placing clear panels on your coop during the winter offers a measure of protection from the elements, but light still filters in and warms up the coop.

A shed roof on this large coop works well and is very attractive too.

air. However, if you are experiencing winds from more than one direction, or continuously shifting wind directions, this may not be feasible.

Another option is to place your ventilation openings high in the coop, where any cold drafts that enter will not be blowing directly on your chickens. You may have already considered this as an option for letting warm air and ammonia gas out of your coop, and if so, then congratulations—you have already taken a major step toward establishing draft-free winter ventilation!

However, even with ventilation openings near the ceiling of your coop, you may still wish to take extra precautions to prevent drafts from entering. Having some form of protective covers for your windows and vents, to shade them from the most direct winds, is one logical option. If your coop has glass windows that open outward, rather than sliding upward, this may be sufficient. For vents, a similarly designed cover made out of wood or metal should suffice.

ROOFING CONSIDERATIONS

Your chicken coop will need a roof—of course—but exactly what type of roof will you choose? Most chicken coops are built using one of two roof styles gable or shed. (You may see

A gable roof is a common roof style for most houses and many chicken coops.

the occasional coop with a gazebo-type roof or some other unusual choice, but the vast majority of coop roofs are gable or shed.)

Not sure what the difference is? A gable roof consists of two sides that come together to form a peak—this is the type of roof commonly seen on houses—while a shed roof is flat and slightly slanted to allow rain and snow to glide off. You might also see flat roofs, which are similar to shed roofs, except that the flat roofs are not slanted.

From a ventilation standpoint, the gable (pitched) roof offers areas under the eaves that are suitable for adding ventilation. From an "ease of construction" standpoint, a shed roof is certainly simpler, so if your construction skills are limited and you want to keep things a bit easier, a shed roof just might be the way to go.

Besides shingles, you could consider an alternative type of roofing material, such as a metal roof. It's an attractive style as well.

Roofing material is another consideration. Will you choose a metal roof? Or opt for shingles? Sometimes your choice will simply be a matter of which materials you happen to have handy. If you have a stack of leftover shingles cluttering up a shelf in the garage, why not give them a purpose on the roof of your coop? Or if you have leftover scrap metal that could be recycled for a coop roof, go for it!

Another option is to purchase polycarbonate corrugated roof panels; these are available at home improvement centers and can be a viable option. Please note, however, that these panels may not be as predator-proof as a metal or shingled roof; they simply don't have quite the same level of security as the more traditional roof options.

4 TOOLS AND NECESSARY SKILLS

By now, you've made a lot of decisions about the coop you're going to build. You've picked your location, figured out the size, and chosen most of the features you want to include. But before you grab a hammer and nails, let's take a look at a few other tools and skills you need to build the coop of your dreams.

In this chapter, we'll talk about different kinds of tools, how to handle them safely, and why you'll probably want to use power tools. We'll also discuss necessary safety gear to keep your project running smoothly.

TOOLS

We'll be honest. It's possible to build your chicken coop using only hand tools, but we don't recommend it. Sure, in some ways there is a certain charm to hand tools and a certain nostalgia to doing things the way they were always done. Hand tools are cheaper, quieter, less expensive. You'll work at a slower pace and your work may, perhaps, benefit for it. Especially if you're planning to build a small coop, working entirely with hand tools could be an achievable option.

Above: Power tools—such as a skill saw and a drill—will make your entire project go a lot faster and easier.

Opposite: It's important to learn how to use the tools needed to build your coop correctly so you can lessen any possible danger to yourself or others. You'll also need to follow all the safety instructions that come with your tools. Doing so will help make your coop-building project a happy and fun experience.

On the other hand, power tools have obvious benefits. You'll save time and plenty of effort. Power tools can roar through tasks that would tax your arms and patience; in the right hands, they can be extremely precise and often do better a job.

The solution for you might be a combination, leaving difficult or time-consuming tasks to the machines and handling the more delicate work yourself. Cost and your own experience level will also play a role in the choice. Here are some tools you'll need to complete the job.

ELECTRIC DRILL

One piece of valuable equipment is an electric drill. While you may find the drill useful on occasion for actually, well, drilling, we're going to hazard to guess that you'll use it far more for a different purpose: driving and removing screws. Like a power saw, it's just one of those tools that can make life so much easier.

Ideally you'll have access to a portable drill that runs on rechargeable batteries. Even if you're going to be doing most of your coop construction near an electrical source, we can't tell you how much more convenient it is to use a battery-powered drill rather than constantly fussing with extension cords. It is preferable to have two batteries. Many electric drill kits come with two, but if yours does not, we recommend buying an extra battery right from the start. This way, you can always have one battery charging while the other is in use—very convenient for those times when your drill starts to fade off-f-f-f-f as the battery dies right in the middle of a task.

One problem with cordless drills is that they can be heavy. The more powerful the drill, the heavier it will be (partly because the batteries

The cordless drill and 6-inch screws will become your friends during this project.

A quality level of sufficient length will be used frequently during the construction of any coop you decide to build. A long level can also double as a ruler for drawing straight pencil lines.

get heavier as they become more powerful). While not conclusive, the voltage of the drill will give you a basic idea of how powerful it is; a 20-volt drill is more powerful than a 14-volt drill. A heavy voltage can make a big difference when you're working all day long or using the drill for very hard work, but for smaller carpentry projects such as a chicken coop, a lighter voltage is probably okay. If you have smaller hands or don't want to struggle with a drill that is too heavy for you, try beginning your search for one around 14 volts. This is a great lightweight size, but it might be a little weak for a coop project. A 20-volt drill will give you plenty of strength, but it is big and heavy. Explore these and also the in-between sizes of 16 and 18 volts.

Another thing to consider when choosing a drill is the capacity of the batteries—this is measured in mAh (milliamp hours), and it helps you know how much electricity the battery can hold. The larger the mAh, the longer your battery will work before needing recharging.

What's another great use for an electric drill? Predrilling nail holes. We've all had it happen: you're hammering along, happy and carefree, when suddenly, your nail turns sideways, or bends in half, or splits the wood, or just plain won't go in. One way to avoid these problems is to use your drill and a narrow to drill tiny holes in the places where you want to nail. Not only will your nails go in easier, but they'll go in straight and limit wood splitting. How great is that?

LEVEL

It's a good idea to own at least one small (24 inches is a handy size) level, as it will help you keep things, well, level. And plumb. Just to set the terms straight in your mind now, remember that **LEVEL** means that something is perfectly "flat" horizontally, while **PLUMB** refers to an object being perfectly straight vertically. It's the same idea both ways, and your level helps you accomplish both these jobs. Keep those bubbles centered!

TAPE MEASURE

There's no doubt about it: carpentry is a task of precision. Even tiny errors in measurement can throw off your building project and cause things to not fit or become crooked. A good tape measure helps you avoid these troubles. There's a famous bit of carpentry advice that you've probably heard, but that we'll throw in here anyway: *Measure twice, cut once.* The point, of course, is that a little extra time spent in preparation is easier and less time-consuming than redoing a task later. However, while you're learning, you may find yourself *measuring three times*, and still goofing up.

Since you'll be using your tape measure often, don't be afraid to spend a little bit extra and get a high-quality tape that will stand up to plenty of use. Tape measures that are 16 feet and 25 feet long are common lengths; either one will probably do fine for your coop project.

A few tools you won't want to be without during this project: a hammer, safety glasses, a nail apron, and a handy carpentry pencil. Take your time and be careful!

In addition to helping you get your cuts right, a tape measure can help your math skills. There's nothing like carpentry work to help you hone your fractions. The tape measures used for carpentry work break inches into sixteenths, which can be a little difficult to work with at first if you aren't used to dealing with small fractions. Take a few minutes to examine your tape measure and really look at the divisions between inches. They have different height markings to help you keep track of different fractions. For instance, the half-inch mark is easy to find—it's halfway (of course) between 0 and 1 inch. It also has the longest mark of any fraction. Next are the two quarter-inch (1/4- and 3/4-inch) marks on either side. Their markings are a bit shorter on the tape measure than the half-inch mark, but still longer than the others. There are four eighth-inch marks (1/8-, 3/8-, 5/8-, and 7/8-inch) stuck in between the quarters and half, with shorter marks still, and last of all are a sprinkling of sixteenth marks—eight of them—which are the smallest marks of all.

What can sometimes be confusing is keeping the equivalents straight in your mind; a half inch is the same as 8/16, etc. But with a little practice—and plenty of measurements—you'll be sure to get the hang of it.

PENCIL

Say, after you get those perfect measurements, how will you remember them? And how will you mark them? You'll need something to write with, and something to mark the places you need to cut. For these jobs, you'll want a carpentry pencil.

Carpentry pencils are tougher and thicker than normal pencils—they have to be, to stand up to heavy use in a construction environment. Why are carpentry pencils flat and not round? It's so they can't roll away and also to provide thicker lead for bolder marks on lumber.

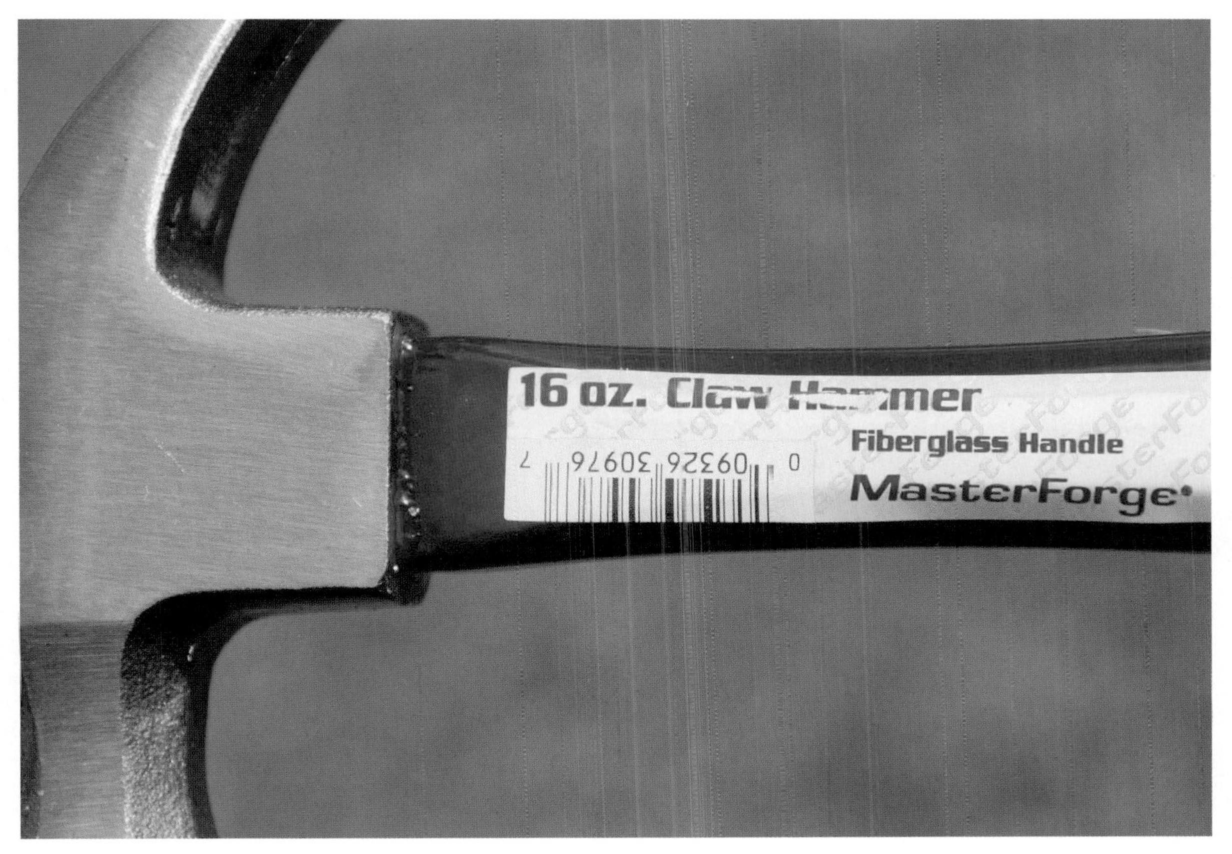

A 16-ounce claw hammer is a great choice for the types of jobs you'll need to do on this project. This hammer is a good weight and can get the work done without being so heavy that it tires your arm.

HAMMER

Nothing says "carpenter" like a hammer. Technically, the type of hammer most people have in mind (and the kind you want) when they say "hammer" is a **CLAW HAMMER**. This is type of hammer where the big blunt end is used for pounding nails; the split back end can be used for removing misplaced or otherwise wayward nails.

You want a hammer heavy enough to do its job well, but not so heavy that you can't handle it. Sixteen- and 20-ounce hammers are common and popular, and the 16-ounce would probably be a good choice for you.

SQUARE

When building a structure such as a chicken coop, carpentry is all about keeping things square. This means making sure that your 90-degree angles are truly 90 degrees (a right angle) and not something slightly crooked such as 88 or 93. If things are square, the walls and roof of your coop will stand straight and proud, and everything will assemble correctly. If things aren't square, you're going to be in trouble (the Leaning Coop of Pisa?).

Luckily the tool you need exists. It's called—get this—a square. (Who said carpentry was difficult?) There are two common kinds of squares: **FRAMING SQUARES**, which are basically large L-shaped rulers, and **SPEED SQUARES**, which are quite a bit smaller and triangle-shaped.

A small (6-inch) speed square is great to keep in your pocket while you work; you can use it for marking cuts and all sorts of jobs. The larger size of a framing square makes it easy to check the squareness of your corners.

RAZOR KNIFE

A razor knife is an item that finds its way into most carpenters' toolkits. You'll use it for all

Make sure that the corners of your coop frame are square and strong by using a framing square. You'll use it (or a similar tool) frequently during the framing process.

You might find it helpful to keep a hand saw around the jobsite—they're fast, simple, and useful for certain jobs.

sorts of little jobs, from trimming shingles and roofing paper to fine-tuning a cut to even sharpening pencils. Take care with it—don't rush—it's very sharp.

HAND SAWS

There are always times—even for professionals—when a hand saw is simply the quickest method for making a cut. In these cases, it's likely because getting out and setting up a power saw would take more time than grabbing your trusty hand saw and taking care of the task by hand.

POWER SAWS

More often, though, you'll find yourself facing a cut—or a series of cuts—that will be very tiring or time consuming, or that require fine precision. In these cases, it is worth your while to consider a power saw for the job.

A common type of household saw—one you probably already have access to—is the well-known **CIRCULAR SAW**. As its name implies, this small saw has a round blade (usually around 6 to 8 inches in diameter—7¼-inch saws are most common), and it is very versatile, able to handle many of the tasks you throw at it. In fact, it's possible to make all the necessary cuts for the coop outline in this book with just a circular saw, so we heartily recommend you obtain one.

Lots of circular saws run off power from their cords, but there are also battery-powered circular saws available. As with a battery-powered drill, the advantage here is being able to move around anywhere, free of cords. While battery-powered saws aren't usually as strong as the plug-in variety, the convenience is valuable.

For the majority of the cuts for your chicken coop project, you'll want to use power saws. The circular saw is one of the most common, versatile, and easy-to-use power saws available.

They are, however, greedy for power, and a cordless circular saw can drain a battery rather quickly. For a chicken coop project (in which you most likely won't be sawing huge pieces of lumber), the lower powers of battery saws are probably okay, but be aware that battery saws are generally more expensive than corded versions.

A **JIGSAW** is another very useful tool. You don't have to have one, but for those times when you're making very small cuts in awkward places, they can be a real time and labor saver. A jigsaw has a blade that moves up and down vertically very quickly, allowing it to be quite maneuverable, albeit slow.

A **CHOP SAW** is also a great time saver, if you have access to one. This saw lets you make very accurate cuts more quickly than a circular saw, and it's also a snap to make cuts of various angles.

SAWHORSES

If you're going to be sawing—and you will—you're going to need somewhere to make your cuts. And if you want to do it right—and you do—you're going to need a pair of **SAWHORSES**. Sawhorses don't need to be fancy, just strong enough to do the job. You can find wooden or metal sawhorses, and even sawhorses made of plastic. Surprisingly, these plastic sawhorses are pretty tough and stable. They're also super light, which makes them easy to move around; plus, they can fold up, which takes up less storage space. So we recommend them. Problems? Plastic sawhorses can't hold very large loads and are somewhat prone to cracking and breaking.

A jigsaw is capable of making precise cuts and will come in handy many times during your coop project, such as when preparing the end of this rafter.

Above: A chop saw is another tool that can greatly speed up your work by allowing you to make precise straight and angled cuts very quickly.

Left: A good pair of sawhorses will be your friend—you'll use them constantly while building your coop. In addition to being helpful for making cuts, a pair of sawhorses can be easily turned into a temporary table.

You'll need a ladder for the "over-your-head" portions of your project. But be sure to use a strong ladder in good condition and follow the ladder's safety instructions. It only takes a few more minutes to do things carefully.

So, what can you do with sawhorses? Plenty. For one thing, obviously, you can make cuts on them. Don't fall for the (easy to make) mistake of cutting your wood *in between* the sawhorses—you'll bind your saw and the whole setup will just fight you. Instead, be sure to use your sawhorses as *supports* for the piece of lumber that you want (usually the longer piece), and let the piece you're cutting away hang off the edge of the sawhorses. In other words, make your cuts *outside* the sawhorses. If it's a long, heavy piece you're cutting, get someone to hold the end so it doesn't fall with a big crash.

Another great thing you can do with sawhorses is create your own temporary worktable. Just use a few thick, sturdy boards, or a sheet of plywood or similar item, on top of your sawhorses to make a big, flat area for you to set tools, make measurements and marks, and work on pieces of your project.

LADDERS

A ladder is essential for this project during the roofing stage. Be sure to use a strong, safe ladder that is in good repair. Safety is the most important thing while using a ladder, especially while working with tools at the same time. Follow the directions carefully (such as which steps you can or cannot climb above) and don't use it on a slope of any kind.

SAFETY ITEMS

It's very important to be conscious about safety when you're working with tools—your chicken coop is not worth an injury! Here are some items to keep on hand.

EYE PROTECTION

Ask any conscientious woodworker about safety, and the topic of eye protection will surely enter the conversation. Any time you're working with nails and hammers, drills and screws, and tools in general, eye protection should be at the front of your thoughts.

A wide range of safety glasses is available; pick a style as elaborate or simple as you like. You can choose from scratch-resistant lenses, lenses that provide UV protection, or even colorful styles. You can even get safety glasses that fit over your ordinary glasses. The style is really up to you. The main thing is to select a pair, and use them. We recommend safety glasses with protection that goes all the way down the sides, as flying bits and stray slivers of wood can easily shoot up while you're working and find their way inside. Don't plan on simply using your ordinary glasses or sunglasses; the glass isn't designed to protect against impacts, and they don't provide protection on the sides. Remember too—especially if you find safety glasses annoying—that you don't have to wear them all the time. Certain jobs (measuring, painting, etc.) don't require glasses, and for the jobs that do, you can always slip your pair on and off between tasks.

EAR PROTECTION

During the loud portions of your project, such as during the use of power tools or hammering, protecting your ears is also important. Sound is measured with the decibel (dB) system, with 0 to 5 representing the faintest flicker of sound a human can hear; 60 decibels represents a normal conservation, and 80–90 represents the level at which hearing damage can occur with prolonged exposure. Since many of the tools used for carpentry are capable of producing noise in excess of 90 dB, ear protection must be used for your safety. There are a few options to choose from, including tight-fitting foam ear plugs or larger headphone-style sets with an emphasis on comfort. Be sure to use your ear protection during the loudest portions of your project.

GLOVES

Gloves are always a tradeoff in situations like this. On one hand (pun intended!), you need protection from rough wood, and the general bangs, bumps, wears, and tears that come with any building project. On the other hand, there will be times during the project where you'll need dexterity and the ability to manipulate small objects. The ideal gloves for you will depend on what's most comfortable to you—you may want a thick pair of very protective gloves for moving wood, digging, and raking, while keeping a smaller, tight-fitting pair around for handling screws and nails or making cuts. Wells Lamont makes a pair of knit gloves with a layer of latex rubber on the palms and fingers, which provides a lot of dexterity while still maintaining excellent protection.

DUST MASK

For people who do a lot of carpentry work, protection from breathing in the sawdust cast up by constant woodcutting is essential. For your chicken coop project, you probably won't be making a large number of cuts, but you're still going to be creating your fair share of sawdust. Once bits of dust become loose in the air, they can be inhaled and potentially cause health problems. For this reason, a simple dust mask can offer inexpensive but effective protection.

Some *very* inexpensive dust masks are only designed to protect against ordinary household dust you might encounter while vacuuming or cleaning a home. For your chicken coop project, we recommend spending a little extra and choosing a sturdy mask.

A lot of people find dust masks uncomfortable, but again, remember that you only have to wear it during the times when you're actually exposed to dusty situations—you don't have to wear it during the whole project.

GET HELP

We'll be honest: you might want some in-person guidance while you build your coop. In the scheme of construction projects, of course, a chicken coop isn't a *major* undertaking, but don't let the small size fool you into thinking there's nothing to it. Unless you already have a carpentry background and experience in building similar projects, we can't stress enough that you should find someone who is experienced and can give you advice and hopefully lend you a hand with the actual construction: someone to show you the safe way to make a cut; someone to show you why *this* or *that* is the best way to do something—someone to lend a helping hand. We'll do all we can in this book, but due to space limitations—not to mention the limitations of the written word—it's always better to have someone with you.

SKILLS

If you were born with a hammer in your hand and could shingle roofs before you could walk, then you can skip this part of the book. If not, read on.

Seriously though, even if you've done carpentry work before, it's never a bad idea to review the basics again before starting such an important project.

NAILING

Nailing can be tricky, but you'll have various reasons to do it during your coop project. To gain practice, why not experiment with pounding some nails into scrap wood first? You'll gain some skill and confidence without risking expensive lumber. Once you've got the hang of it, there are a few pointers to keep in mind.

First, always be careful not to insert a nail too close to the edge of a piece of wood. Depending on the type of wood and the style of nail, nailing too close to an edge can cause the wood to split—a very unpleasant thing because it's unattractive and weakens your wood.

You'll also want to avoid nailing into knots in the wood. Knots are those circular, darker areas of the wood (they were originally the locations of branches in the tree) and they are very hard—much harder than the surrounding wood. Ask any carpenter: trying to drive a nail through a knot is just asking for trouble.

If you have access to an air nailer (and, hopefully, someone experienced in using it safely and effectively), you'll be amazed at how fast you can fly through your project, especially when shingling or constructing a nest box. You'll be able to almost effortlessly fasten one nail after another, even in tricky situations. Air nailers can be somewhat expensive and do

There will be times during your coop's construction when it will be a good idea to have someone with carpentry experience help you out with the more difficult or challenging aspects of the project.

Becoming skilled and comfortable with your hammer simply takes practice. Luckily, your chicken coop project will give you ample opportunity to do so.

require an air compressor to work, and as with all power tools, safety must be a top priority at all times. So even though you can rely on a hand hammer and your trusty power drill for screws, if you have the opportunity to use an air nailer on your project, it's not a bad idea.

PULLING NAILS

It happens to everyone: you've put a nail somewhere you didn't want it, and now it needs to come out. Or, you're in the process of pounding a nail and it bends or twists beyond saving. Use the back end of your claw hammer

An air nailer isn't an essential tool, but it's one that can really save you a lot of time and effort—especially when shingling.

to help leverage the nail back out; a side-to-side motion is usually more successful the pulling straight out. You might also need a block of scrap wood to use as a fulcrum. And wear those safety glasses!

SCREWING

In this book, we recommend that you use screws and a power drill to fasten your coop's basic frame. There are a number of reasons why. For one thing, screwing with a power drill (especially long screws) is generally easier than pounding nail after nail—especially for beginners who may not have a lot of experience with extensive hammer use. We would even propose that screwing (again, for beginners) is more accurate and faster than nailing.

Using screws and power drills is generally safe, but there are some tips and safety suggestions that will make your job easier. When starting a screw into the wood, use a very slow speed on the drill until the screw has worked its way into the wood a short distance; you can easily adjust the speed of the drill with the trigger. Once your screw is in far enough for it to stand on its own, release it and use both hands to hold the drill—this will give you more control of the tool and keep your hands out of the way in case your drill slips off the screw. Then you can feel free to speed up and finish the screw at a more rapid pace. Don't worry if your first attempts aren't too fast—you'll quickly get the hang of this and be inserting screw after screw in no time.

RIPPING

RIPPING is the term used for cutting a piece of wood along with the grain—that is, parallel to the grain. This is in contrast to a **CROSS CUT** (the kind you likely think of when you imagine someone sawing wood), which goes sideways across the grain—perpendicular to it. Ripping comes in handy regularly, and you'll use it for a few different cuts during your coop project.

The problem is that ripping can be a little more challenging than your regular cross cut. The easiest way to accomplish it is to use a **TABLE SAW**. A table saw works and looks a bit like a miniature sawmill—long pieces of wood can easily slide across the blade and be ripped with ease and accuracy. Unfortunately, table saws are big and expensive, and usually not part of the average person's tool inventory. You might know someone with a table saw—maybe Uncle Joe has a woodworking hobby and lots of cool tools—but then again, it can become cumbersome as soon as you start having to move lumber back and forth. Luckily, your trusty circular saw can make a rip cut, but you may need to fashion a guide to help you keep the

Screwing is a skill that you can learn pretty fast. Take your time in the beginning while you get the hang of it, and you'll be sailing through that box of screws before long.

It takes some experience to rip wood properly and safely, so this is a great area to get some help and advice if you're unsure. You won't have to make very many rip cuts during this project.

cut straight, or at least proceed quite slowly and carefully. Another saw that can make rip cuts is a **BAND SAW**, but again, it's not usually a stock piece of most home tool kits.

GLUING

Sure, you're going to fasten everything with screws or nails, but applying a bit of wood glue to each joint will help keep things tight and sealed for years to come. It's just a little backup plan. You don't even have to wait for the glue to dry—you just apply it to the surfaces you're about to join and then immediately fasten them with your screws or nails. Don't go crazy with wood glue, but do use it liberally—enough of it to get the job done right.

Get used to fastening everything with screws or nails, but a handy bottle of wood glue can help keep joints snug and tight. Glue is easy to apply and only adds a few extra minutes.

Load up your tools, get your materials, and let's get going! The coop awaits.

5 BUILDING THE COOP

At last! The time has come! You've considered the advantages, you've chosen a location, you've decided which type of coop you need, and you've gathered the necessary tools for the job. Now it's time to get started on building your coop.

We've provided instructions for a basic 4 × 8-foot coop that is suitable for four chickens. We'll show you—step by step—how to build the coop, and we've included plans, a materials list, and a cut list so you'll have everything you need to get going on your coop-building project.

The coop we'll build in this chapter is an attractive, strong, durable structure, yet one that is still small enough for a home project. We've used heavy-duty, high-quality materials, including cedar on all exposed elements for durability and resistance to rot. Because we live in a northern region with very cold winters, this coop is designed with thick walls to help insulate against cold and wind. All in all, the materials for this coop cost $680 at our location at the time of this writing.

For easy everyday cleanout, there is a small exterior access door that allows you to reach into the coop for cleaning without having to go inside through the main door. There is also another small exterior door that allows access to the nest boxes—again, so you won't have to go in with your chickens every time you want to collect eggs (this is a great convenience and time saver). The spacious run will give your chickens room to move around and scratch. For complete access to the coop's interior, there is a removable coop wall, and while it's not something you'll be removing daily, it is still simple to remove for times when you want to easily access your coop's interior for thorough cleanings. Finally, there is a large main door giving your chickens convenient access to the run.

Let's go!

It's building time! By following the directions in this chapter, you'll be able to build a high-quality coop that is large enough for four chickens. The coop is equipped with doors for easy cleanout and egg collecting, a built-in run, and an attractive design.

BUILD INSTRUCTIONS

COMPLETE CUT, HARDWARE, AND MATERIALS LIST

Below is a full list of lumber, hardware, and materials you will need to build this chicken coop. Individual steps of the building process break the list down further.

CUT LIST

Furring strips

- 2—14 3/8" 1 × 1 1/2" furring strips
- 2—16" 1 × 1 1/2" furring strips
- Total length needed: 3'

1 × 4s

- 1—34" cedar 1 × 4
- 2—62" cedar 1 × 4
- Total length needed: 13'

1 × 6s

- 2—96" cedar 1 × 6 (ripped 40°)
- Total length needed: 16'

1 × 12s

- 1—12 7/8" cedar 1 × 12
- 1—54" cedar 1 × 12
- 2—30 1/4" cedar 1 × 12
- Total length needed: 11'

2 × 2s

- 2—96" cedar 2 × 2
- 4—34 1/2" cedar 2 × 2 (one end cut 40°)
- Total length needed: 28'

2 × 4s

- 3—21" cedar 2 × 4
- 2—59 3/8" cedar 2 × 4
- 4—45" cedar 2 × 4
- 10—59 3/4" cedar 2 × 4
- 2—96" cedar 2 × 4
- 2—30 3/8" cedar 2 × 4
- 8—31 1/2" cedar 2 × 4
- 2—12" cedar 2 × 4
- 1—8 1/2" cedar 2 × 4
- 1—8 1/4" cedar 2 × 4
- Total length needed: 126'

2 × 10s

- 12—35" cedar 2 × 10
- 7—42" cedar 2 × 10
- 1—44 1/4" cedar 2 × 10
- Total length needed: 64'

2 × 12s

- 4—96" cedar 2 × 12
- 2—96" cedar 2 × 12 (one side ripped 40°)
- 2—10 3/4" cedar 2 × 12
- 3—45" cedar 2 × 12
- Total length needed: 62'

Plywood

- 1—44 1/4" × 44 1/4" 3/4" plywood
- 1—33 1/8" × 13 1/2" 1/2" plywood
- 2—33 3/8" × 3" 1/2" plywood
- 2—14" × 18 1/4" 1/2" plywood

Roosts

- 1—25" roost
- 1—40" roost
- 1—33 1/4" roost

Cedar slats

- 13—1/2" × 3/4" cedar slats

HARDWARE AND MATERIALS LIST

- 4—4" L braces
- 1—door handle
- 3—hook-and-eye latches
- 1—3" pair of hinges with removable pins
- 1—8" × 8" gable vent
- 2—2" hinges
- 2—hook latches with eyes
- 4—2½" hinges
- 2—latches

Hardware cloth

- 1—56" × 24" hardware cloth
- 2—64" × 59" hardware cloth
- 1—25" × 19" × 19" Isosceles triangle, hardware cloth
- 1—41" × 27" × 27" Isosceles triangle, hardware cloth
- 2—3" × 64" hardware cloth

Additional materials

- 2—100" × 36" Sections of roofing paper
- About 5 bundles cedar shingles

Fasteners

- wood glue
- roofing nails
- staples (for staple hammer)
- 6" exterior screws
- 3" exterior screws
- 3" nails (air nailer or hand nails)
- 2" nails for shingles (air nailer or hand nails)

STEP 1: THE DOOR

Cut list for the door:

- 3—21" cedar 2 × 4s
- 2—59³/₈" cedar 2 × 4s

Hardware:

- 4—4" L braces
- 1—handle
- 1—latch

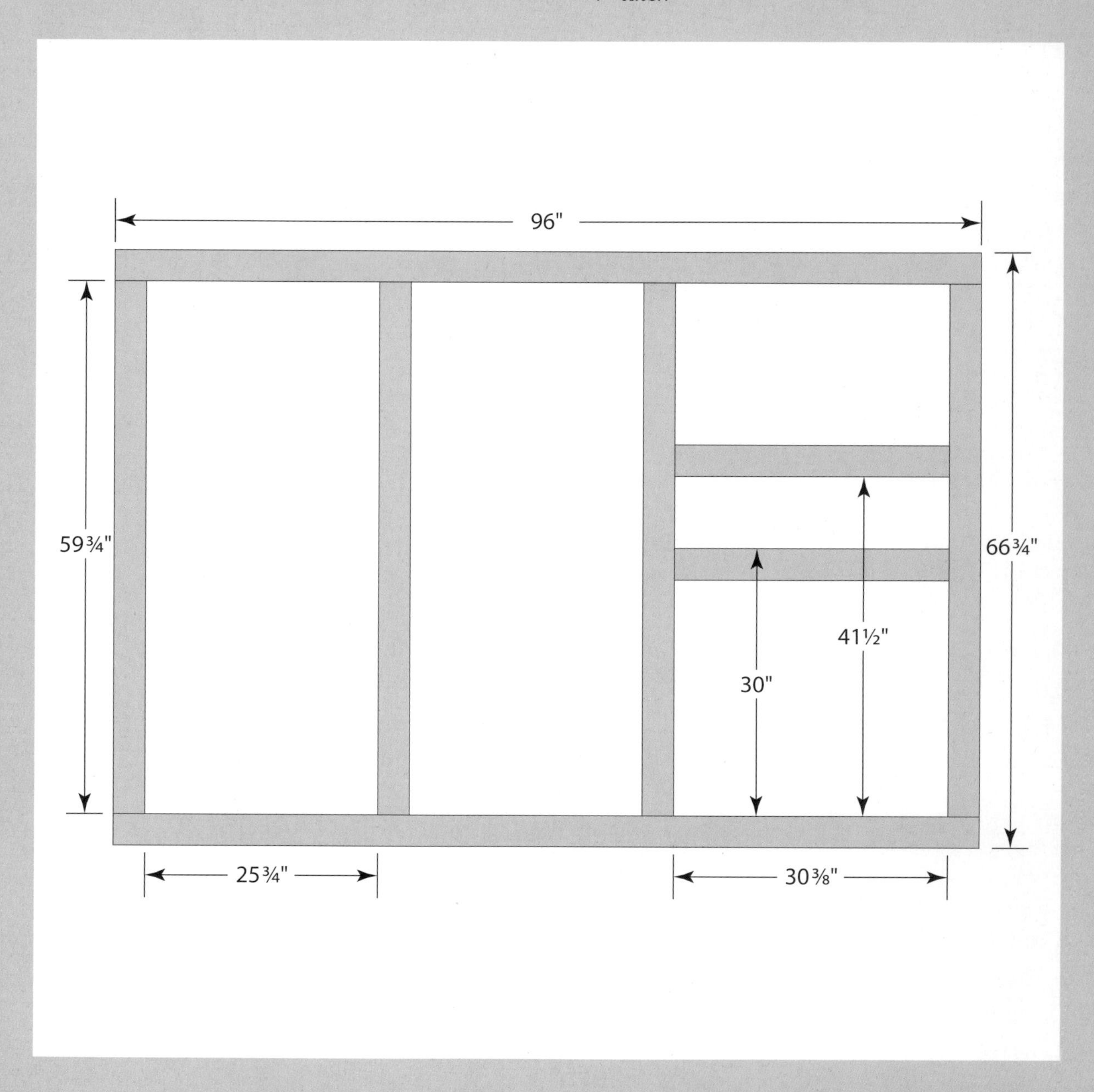

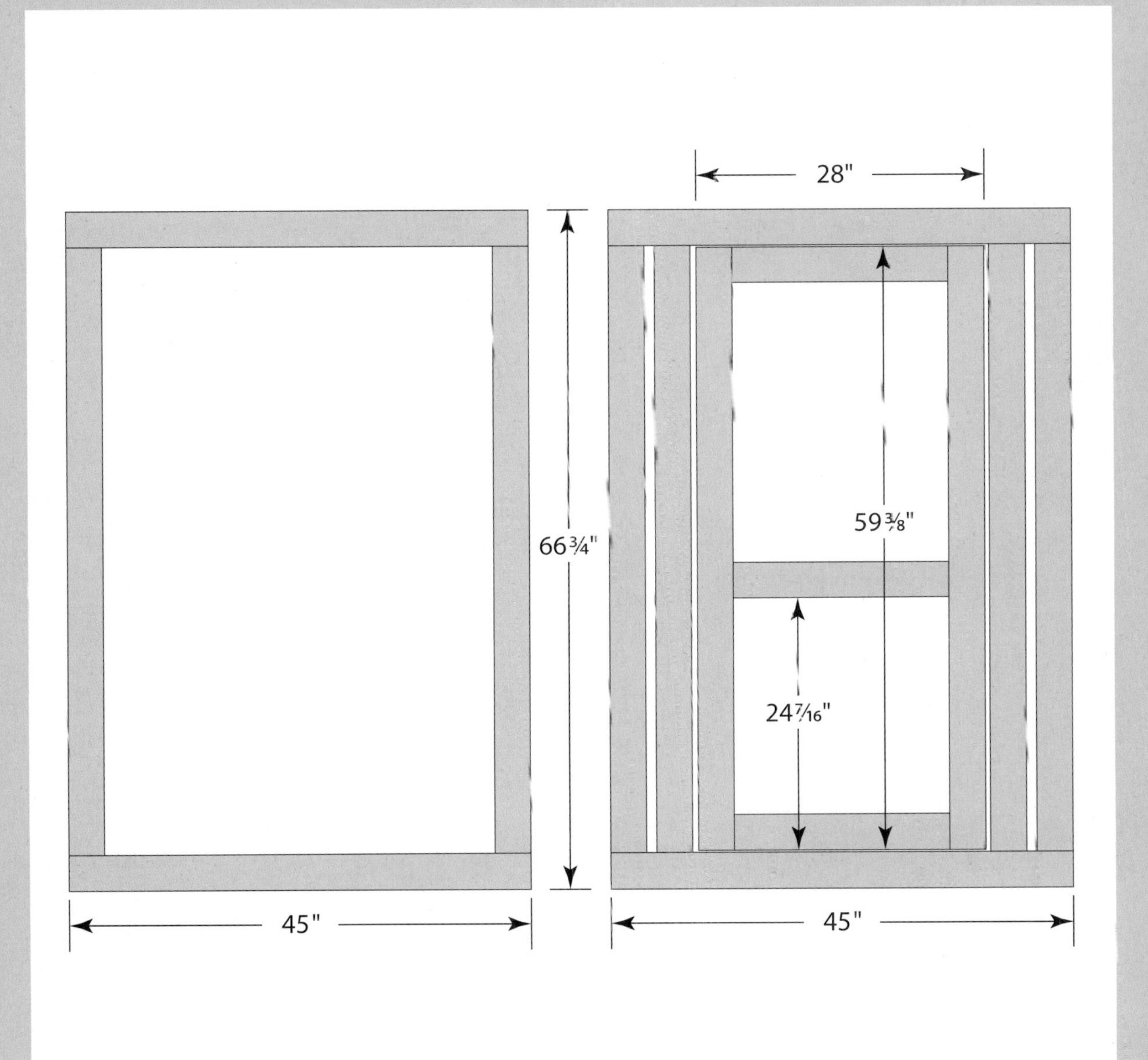
28"
66¾"
59⅜"
24 7⁄16"
45"
45"

BUILD INSTRUCTIONS (CONTINUED)

Let's start out by building something fun: the door of the coop. It's relatively simple. The door is constructed out of two 59³/₈-inch 2 × 4s for the long sides and two 21-inch 2 × 4s for the top and bottom. There's also an additional 21-inch 2 × 4 set in the middle of the doorframe, 24⁷/₁₆ inches from the bottom of the door. Use 6-inch screws to fasten, along with a bit of wood glue.

Your coop's door will see a lot of use and wear over time, so for additional strength, add a 4-inch L brace at each corner to further anchor the joints.

Your door will use four L braces on the joints to help give it strength. It's helpful if you hold the bracket in the correct position first and make marks for your screw holes with a pencil.

The completed door, with L braces installed. At this point, we haven't installed hardware cloth on the door.

STEP 2: THE FRONT PANEL FRAME

Cut list for the front panel frame:

- 2—45" cedar 2 × 4s
- 4—59¾" cedar 2 × 4s

Hardware:

- 1—pair of 3" hinges with removable pins

When working on your coop's frame, it's helpful to use a framing square to help make sure your corners are straight. Here, we're using a framing square on the front panel frame.

Another useful tip when squaring the frames is to measure the diagonals and make adjustments to the frame until both diagonal measurements match. Leave one corner unfastened until you are square: this will give you enough flexibility to make adjustments.

BUILD INSTRUCTIONS (CONTINUED)

Your coop's front panel consists of a basic frame constructed of two 59¾-inch 2 × 4s for the sides and two 45-inch 2 × 4s for the top and bottom. You'll also need an additional pair of 59¾-inch 2 × 4s, located just inside the outer 59¾-inch 2 × 4s. Make a 1¼-inch gap between the two, for a total span of 8¼ inches from the outer edges of both. Fasten all joints with 6-inch screws and wood glue.

Now is as good a time as any to attach the door to its frame, so use your two 3-inch hinges to perform this task.

When positioning and attaching the door to the front panel, try using nails at each corner as spacers. This will keep the door from binding against the frame.

With your spacers in place, go ahead and screw in the hinges to attach the door.

BUILD INSTRUCTIONS (CONTINUED)

STEP 3: THE BACK PANEL

Cut list for the back panel:

- 2—45" cedar 2 × 4s
- 2—59¾" cedar 2 × 4s

The back panel couldn't be any simpler: use the two 59¾-inch 2 × 4s to create the sides, and tie that together with the pair of 45-inch 2 × 4s for the top and bottom. Fasten with 6-inch screws and wood glue.

The front panel (with the door) and the back panel are now complete.

STEP 4: THE SIDE PANELS (×2)

Cut list for the side panels:

- 2—96" cedar 2 × 4s
- 4—$59\frac{3}{4}$" cedar 2 × 4s
- 2—$30\frac{3}{8}$" cedar 2 × 4s

Things get slightly more complicated with the larger side panels, but it's still very manageable. Use the two 96-inch 2 × 4s to create the top and bottom, along with a pair of $59\frac{3}{4}$-inch 2 × 4s for the sides. A third $59\frac{3}{4}$-inch 2 × 4 should be placed parallel to the left side of the panel, with a $25\frac{3}{4}$-inch gap separating this third $59\frac{3}{4}$-inch 2 × 4 from the side board. The fourth $59\frac{3}{4}$-inch 2 × 4 is placed (again) $25\frac{3}{4}$ inches from the third 2 × 4—this will leave a gap of $30\frac{3}{8}$ inches between the fourth 2 × 4 and the right edge of the side panel. This $30\frac{1}{2}$-inch area will be the location of the actual coop portion of your project.

Within this $30\frac{3}{8}$-inch gap is where those two $30\frac{3}{8}$-inch 2 × 4s should be located. They should be placed parallel to the top and bottom of the frame, with the top of the lower $30\frac{3}{8}$-inch 2 × 4 spaced 30 inches above the top of the lower 96-inch 2 × 4.

The second $30\frac{3}{8}$-inch 2 × 4 should be placed above the first $30\frac{3}{8}$-inch 2 × 4, so that there is an $11\frac{1}{2}$-inch gap between them.

Repeat the above steps to create a second side panel that is identical to the first.

A mostly completed side panel with hardware cloth already in place. You'll need two of these side panels (they're identical). Note that during this coop build, we added the $30\frac{3}{8}$-inch 2 × 4s later on in the process, so you don't see them in this photo. But we recommend adding them at this step.

BUILD INSTRUCTIONS (CONTINUED)

STEP 5: HARDWARE CLOTH

Hardware:

- Hardware cloth
- Roofing nails

You now have the frames for all four sides of your coop. A good thing to do at this point—while they're still separate from one another and easier to handle—is to add the hardware cloth.

For the door, you will need to cut a piece of rectangular hardware cloth that is about 56 inches on the long side and 24 inches on the short side. These numbers do not need to be exact, but you'll want to be close.

For each side panel, you'll need a larger piece (or probably two pieces with a seam)—enough to cover about 64 inches on the long end by almost 59 inches on the short edge. Note that you only need to cover about two-thirds of each side panel—you don't need to cover the section of each side panel with the two 30³/₈-inch 2 × 4s, as this area will be covered by wood later on.

With someone to help you, carefully roll out the hardware cloth to cover the door and the required sections on the side panels. Gloves and glasses are essential here, along with a pair of tin snips to cut the hardware cloth to size. Cutting wire is unpredictable, so have a helper for this step to avoid injury.

The hardware cloth can be fastened with roofing nails. In these images we're using 1/4-inch hardware cloth; 1/2-inch is also commonly used.

You must be very careful when working with hardware cloth—it's sharp, and it has a tendency to "jump" back into its original shape. Make sure you've got your gloves! This is one area where you'll definitely want another pair of hands to help you. One way to tame the wild hardware cloth beast is to fasten two top corners first, then work your way down. Don't fasten all four corners at once; you'll need to work in one direction so you can keep stretching it taut and avoid ending up with slack in the middle. Use roofing nails to fasten; the large heads will keep the mesh in place and the short length makes for easy installation.

STEP 6: COMBINE THE FRAME PANELS

This is a fun step because you get to make some visible progress in a hurry. This is where your coop goes from looking like just a pile of lumber to a real structure. Using our old friends the 6-inch screws and wood glue, attach one of your side panels to the back panel (make sure you attach the back panel to the correct end of the side panel—the end with the wider 30³/₈-inch gap and the two 30³/₈-inch 2 × 4s). You'll need some extra hands around for this portion of the project. You'll also want to work in a fairly level place.

With the hardware cloth in place, it's time to assemble the coop's frame. Note that the front and back panels need to go on the "inside" of the side panels.

Apply some wood glue to the frames before joining them together.

Be aware that your long side panels go on the "outside," and your shorter front and back panels fit in the "inside" of those.

Once you have a side panel and the back panel fastened, and they are standing on their own, you are free to add the other side panel, as well as your front panel. Combine all these together with 6-inch screws and wood glue.

This is where your framing square comes in very handy, as you'll use it repeatedly to ensure that your corners are square. Check before, during, and after the fastening process.

While you're at this stage, there's another trick you can use to double check that things are square. Use your tape measure to measure across the *diagonal* of your frame—both ways. Are they the same? Then you're good to go! If not (they're probably very close), then you're going to have to try to bend and shift your frame a little bit one way or the other until you get essentially the same measurement on both diagonals.

When assembling the frames, start your screws ahead of time as shown here. Then when you're ready and holding everything in place, you can quickly and easily finish the screws off without having to worry about getting them in the right places.

Your framing square can come in handy again during this stage. You want all four walls to be square in order to avoid creating bigger problems later on.

Don't try to handle the frames alone. They're awkward and heavy and you'll save time and be more accurate if you have another pair (or pairs) of hands around. Here, the final side panel is being attached.

BUILD INSTRUCTIONS (CONTINUED)

Take care to keep things flush while screwing the frames together. If you're working in grass or another type of slightly uneven surface, you can temporarily place a spare piece of lumber under your frames to help keep them steady and level.

With all four sides up and fastened, use your tape measure once more to check the diagonals. Again, having someone to help you will ease the process.

At this stage, we have all four panels built and assembled, with hardware cloth on the necessary areas. Time to move on to the roof!

STEP 7: THE RAFTERS

Cut list for the rafters:

- 8—31½" cedar 2 × 4s

This coop is designed with four rafters, and they're fairly straightforward to construct. Make a cut at a 40-degree angle to the short end of all eight 31½-inch 2 × 4s. Then, cut a 3¼-inch × 1½-inch notch into the opposite end of just four of the 31½-inch 2 × 4s. This notch will let the rafters fit snugly onto the top of the side panels and give them plenty of strength. Be sure to refer to the diagrams while making these cuts.

Next, take four pairs of 31½-inch 2 × 4s and combine them at the ends with the 40-degree angle cuts (use your trusty 6-inch screws for this job as well). When you're done, you'll have four rafters completed and ready to screw (with 6-inch screws) onto the coop frame. The two unnotched rafters go on the front and back ends, while the two notched rafters are installed directly above the 59¾-inch 2 × 4s along the side panels.

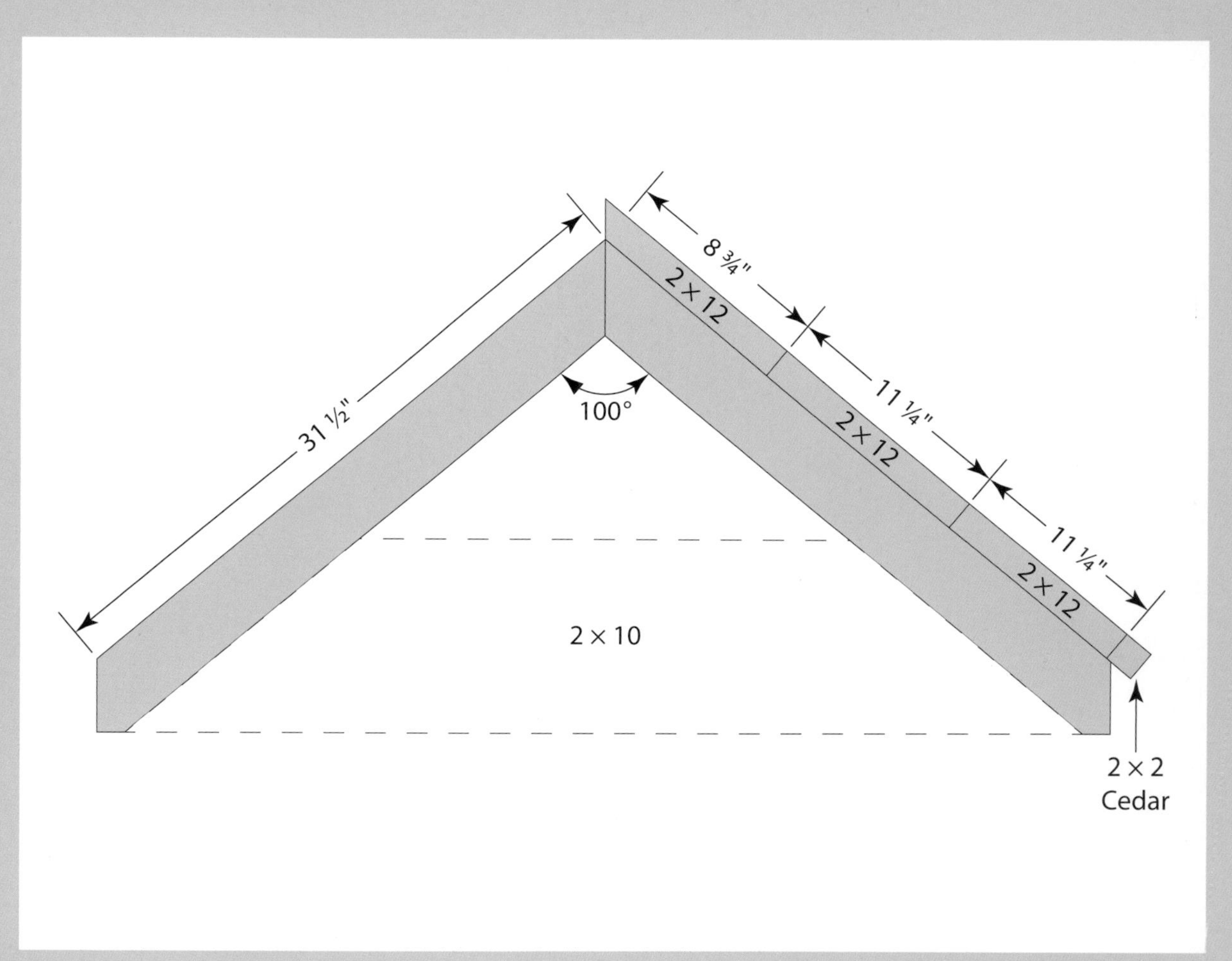

Use 6-inch screws to fasten your rafters to the frame. It's important to keep the rafters straight (square to the frame) as you work.

The coop uses four rafters: two (unnotched) over the front and back panels and two (notched) above each stud.

This photo shows the orientation of the notches that are needed on the two middle rafter sets.

STEP 8: THE ROOF BOARDS

Cut list for the roof boards:

- 4—8' cedar 2 × 12s
- 2—8' cedar 2 × 12s, one side ripped 40°

Now you can put up the 2 × 12 roof boards. Don't forget to rip the two ridge boards to 40 degrees.

Nail your roof boards into all four rafters, either by hand or by using an air nailer.

Now, let's get a roof to go with those rafters. With someone to help you (those 2 × 12s can be heavy), install all six 2 × 12s to the rafters with nails. Use three 2 × 12s on each side of the coop, saving the two that you ripped to 40-degrees for the tops.

STEP 9: ROOF TRIM

Cut list for the roof trim:

- 2—96" cedar 2 × 2s
- 4—34½" cedar 2 × 2s (one end cut 40°)

To keep things looking nice and tidy, add some trim to the roof. Use the two 96-inch 2 × 2s and attach them with 3-inch screws directly onto the lowermost 2 × 12 on each side of your coop's roof. The four 34½-inch 2 × 2s are used as trim on the front and back ends of your coop, spanning all three 2 × 12s on each side and meeting at the top with the 40-degree cuts. These are also attached with 3-inch screws.

Roof trim will give your coop a professional touch. You'll need a total of six pieces of trim for the roof. Two for each side . . .

. . . and two apiece for the front and back. Use 3-inch screws.

STEP 10: ADD ROOFING PAPER

It's best to put something in between your 2 × 12s and your shingles. Felt roofing paper is easy and quick to install and will help keep moisture out. Cut a piece for each side of the roof to roughly the size you need (100 × 36 inches); you can trim it to fit using your razor knife later.

Attaching the roofing paper goes fast; we used a staple hammer to quickly tack the felt into place, but you can also use roofing nails and a hammer. Whatever you have will work just fine.

Roofing paper comes next, for both sides of the roof. Overlap the paper across the ridge to provide added protection against moisture.

You can tack down your roofing paper with roofing nails, or—for a faster and easier solution—use a staple hammer. It's a great tool.

BUILD INSTRUCTIONS (CONTINUED)

STEP 11: THE SIDE WALLS

Cut list for the side walls:

- 12—35" cedar 2 × 10s

Let's get some walls on this coop. Working your way up from the ground (and on the *inside* of the coop), use nails to attach three of your 35-inch 2 × 10s to the lower portions of each side panel. This will be enough to cover the section that's just below each of the lowermost 30³/₈-inch 2 × 4s present on each side panel. So you'll use a total of six 35-inch 2 × 10s to cover the lower areas of both side panels.

Next, use your remaining 35-inch 2 × 10s to cover the upper areas of both side panels—the areas above the upper 30³/₈-inch 2 × 4s (the open space left over will become your cleanout and nest box doors). For each of the uppermost 35-inch 2 × 10s, you'll need to cut a notch 5¹/₂ inches long and 1¹/₂ inches wide to fit around the rafter.

Installing the coop's side walls shouldn't take long, and it can be a fun part of the project as your coop begins to really take shape.

The back wall is also pretty simple—just seven 2 × 10s stacked on top of each other. Aren't you glad you were careful to keep everything square?

An in-process view of the construction of the back and side walls.

STEP 12: BACK WALL

Cut list for the back wall:

- 7—42" cedar 2 × 10s

The back wall is simple. Just use all seven of your 42-inch 2 × 10s to fill in the back panel. As with the side panel walls, you'll be doing this from the inside. Fasten with nails, and you're good to go.

STEP 13: SHINGLES

Materials list:

- Roughly 5 bundles of cedar shingles

At this point, you can go ahead and shingle the roof, side panel walls, and back wall of the coop. There are many different routes you can take here, but this plan calls for the use of cedar shingles. In all cases (whether roof, side wall, or back wall), start shingling from the bottom and work your way up. This way, the shingles will overlap in a way that helps prevent water from penetrating the coop. Using a carpenter's chalk line or a long level and pencil can help keep your lines straight, although you may

During the shingling process, you'll often need to trim or otherwise adjust certain shingles to make them fit, particularly when you reach the end of a row. A pair of sawhorses and a handsaw can make quick work of this task.

With one row of roof shingles complete, it's time to start overlapping with the second row.

Install shingles on your coop's roof, side, and back panels. You can use a hammer and nails to fasten your shingles or an air nailer if you have access to one. (Be careful!)

BUILD INSTRUCTIONS (CONTINUED)

This is a side view of the way the roof shingles overlap after completion.

If you do decide to try an air nailer for the shingling process, be sure to read the manual and fully understand how the tool works, including its safety precautions. Take your time, be careful, and wear safety glasses.

Sometimes you'll want help from an experienced person for challenging parts of the construction.

prefer to stagger the shingles for a more rustic look. In any event, nail your shingles either by hand or by using an air nailer (this can greatly speed up and ease the process). Also, keep a hand saw and razor knife handy to trim any shingles to size as required.

When you finish each side of the roof, you can use a circular saw to trim off whatever excess shingles extend over the edge.

STEP 14: RIDGE CAP

Cut list for the ridge cap:

- 2—8' cedar 1 × 6s, ripped 40°

After your roof is shingled, you'll want to put a professional finish on the ridge. Use two 8-foot (96-inch) 1 × 6s, ripped to 40 degrees to handle this job. Nail them in place.

Once the roof shingles are finished, you can top it all off with a ridge cap. Don't forget to make those 40-degree rips!

STEP 15: BUILD GABLE VENT AREA

Cut list for the gable vent area:

- 2—10¾" cedar 2 × 12s or 2 × 10s
- 2—12" cedar 2 × 4s
- 1—8½" cedar 2 × 4
- 1—8¼" cedar 2 × 4

This coop includes an 8 × 8-inch gable vent to help with ventilation inside the coop (as discussed in Chapter 3). We're going to place this vent on the back gable, but before we do that, we need somewhere to attach it.

We need to make two right triangles out of the 10¾ 2 × 12s. The two "straight" legs (that meet to form the right angle) need to be 10¾ inches long and 9 inches high. This will require making several cuts into your boards.

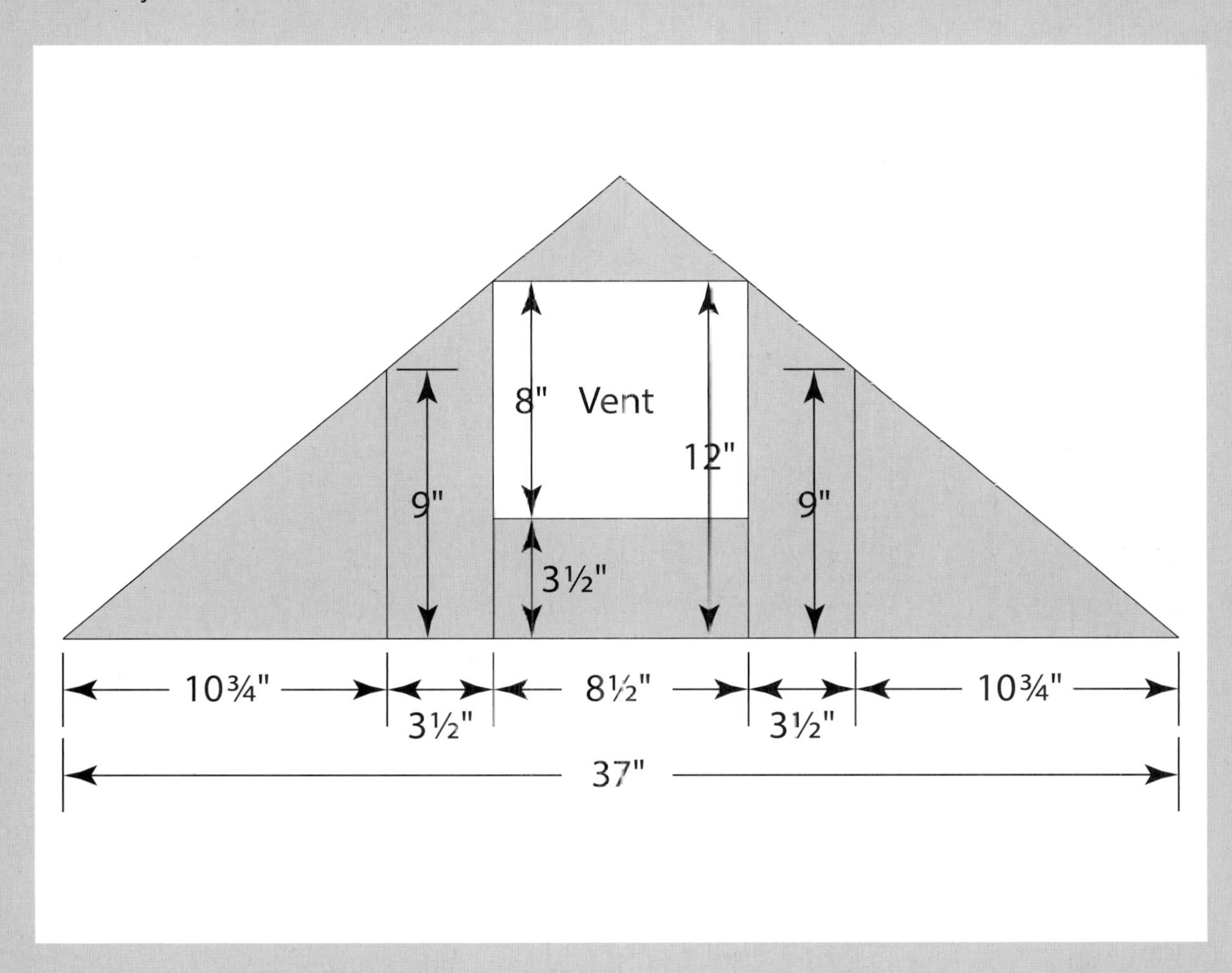

BUILD INSTRUCTIONS (CONTINUED)

Your 12-inch 2 × 4s need an angled cut as well. Make a mark at 9 inches on one side, and then make a diagonal cut from this point to the corner of the board, as shown in the diagrams. If you stand your finished board upright on the flat edge opposite the cut, the board should have a slanted "roof," with the highest point being 12 inches from the base and the lowest point being 9 inches from the base.

The $8\frac{1}{4}$-inch 2 × 4 will be used at the very peak of your gable, fitting in just beneath the rafter. As a result, it will require a pair of 40-degree cuts in order to shape it into a triangle. The $8\frac{1}{2}$-inch 2 × 4 will be used at the center of the bottom of the gable to fill in that portion.

All of this will leave you with an 8 × 8-inch hole, just right and ready for your gable vent. Nail it in. With the gable vent installed, you can go ahead and shingle this portion of the coop as well.

This is what the gable vent area will look like from the inside of the coop when it is completed. The 8 × 8-inch vent provides additional ventilation.

STEP 16: COOP FLOOR

Cut list for the coop floor:

- 3—45" cedar 2 × 12s

Building your coop's floor is very simple. It consists of just three 45-inch 2 × 12s placed horizontally on top of the side panels' lower six 2 × 10s. The floorboards will fit right into place and can be nailed down to lock them into position.

The 2 × 12s that make up the coop's floor sit nicely on top of the coop walls.

STEP 17: UPPER VENTED AREA

Cut list for the upper vented area:

- 1—44¼" cedar 2 × 10

Hardware:

- Hardware cloth

Now we can begin to construct the front wall of the coop area. Make two 40-degree cuts on both ends of your 44¼-inch 2 × 10. When you're finished, you should have a trapezoid 44¼ inches long on the bottom and 22½ inches on the top. Nail this just underneath the second rafter from the rear of the coop. It should fit snugly into place, leaving a triangular gap above it, below the rafter. Cover this triangular area with hardware cloth.

This upper area provides additional ventilation. The 44-inch 2 × 10 fits right under a rafter, and hardware cloth is used to protect the integrity of the coop.

STEP 18: ROOSTS

Cut list for the roosts:

- 1—25" roost
- 1—40" roost
- 1—33¼" roost

It's important to give your birds plenty of roosting areas, and as discussed in Chapter 3, these roosts are ideally wooden and somewhat rounded. There are various ways to achieve this; we used a type of hand railing that seemed to have just the right amount of roundness without being too slippery. In any event, make a 45-degree cut on each end of the 25- and 40-inch roosts. Attach these to the left corner of your coop's interior with 3-inch screws, with the shorter 25-inch roost placed higher than the 40-inch perch. The exact height of each roost isn't critical, but aim for about 24 inches and 30 inches off the coop floor. Don't make 45-degree cuts on the 33¼-inch roost, as it is going to be placed just beneath the 8 × 8-inch vent and parallel to the coop walls.

Three roosts arranged like this provide plenty of roosting room for your chickens.

BUILD INSTRUCTIONS (CONTINUED)

STEP 19: REMOVABLE COOP WALL

Cut list for the removable coop wall:

- 1—44¼" × 44¼" plywood, ¾" thickness (for wall)
- 1—12⅞" cedar 1 × 12 (for door)
- 2—14⅜" 1" × 1½" cedar (for door trim)
- 2—16" 1" × 1½" cedar (for door trim)

Hardware:

- hook latch (for door)
- hook latch eyes (for door)
- pair of 2" hinges (for door)

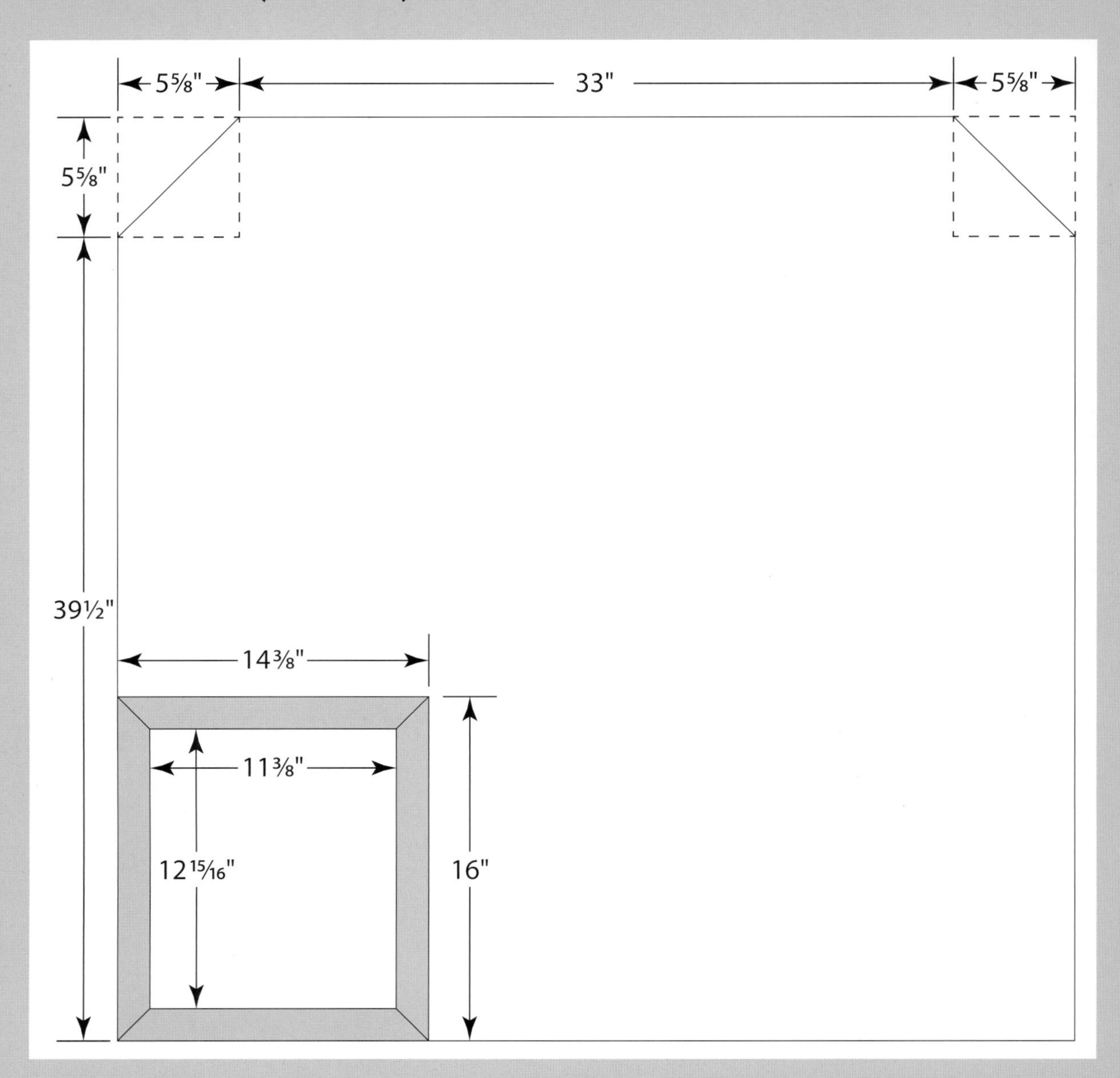

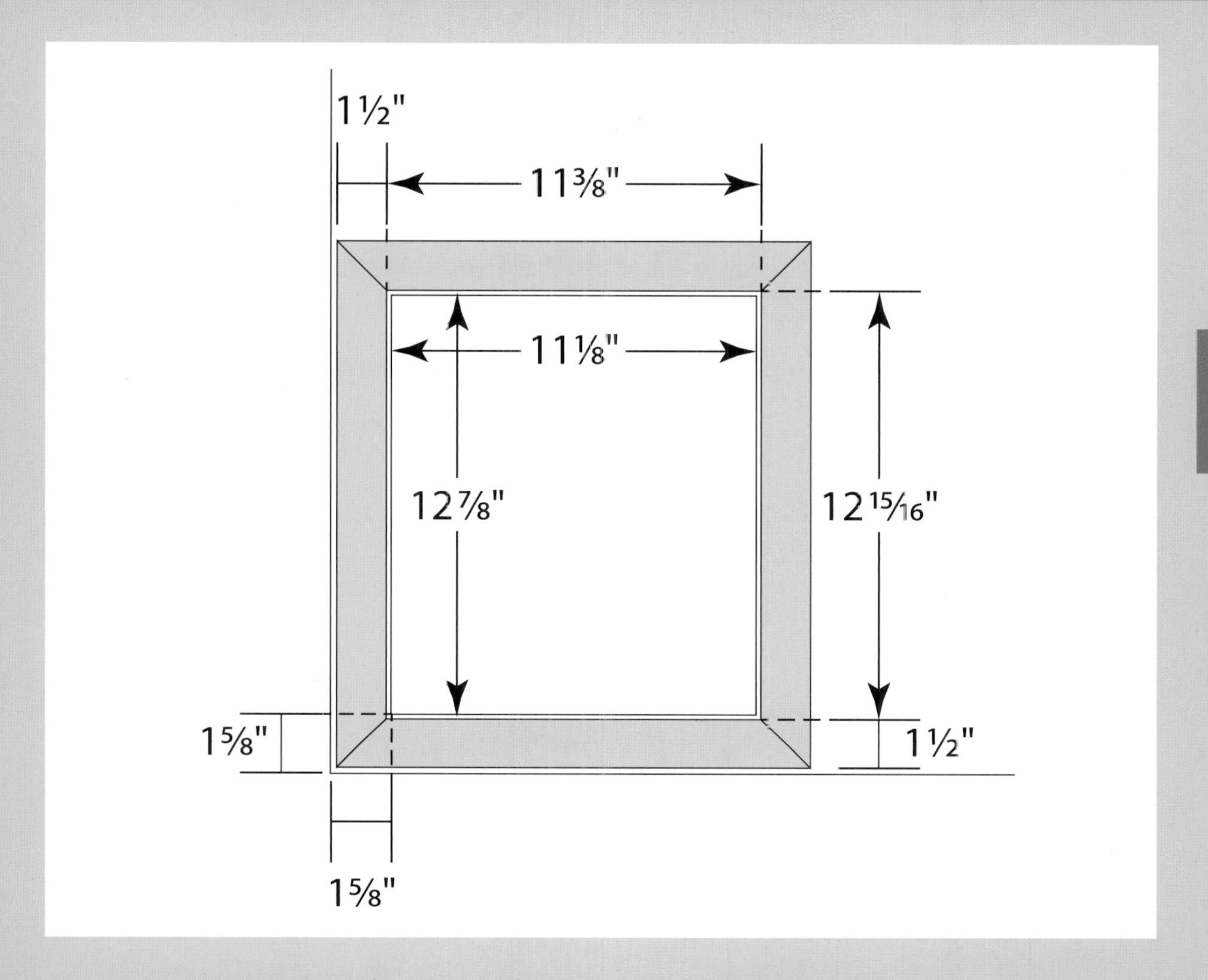

We're really getting close now! The next step is to build the interior coop wall, which is semi-removable. You won't be taking it off daily, but for times when you'd like to have complete access to the coop (as you would for a thorough cleaning), removing the wall is easy enough. But for right now, let's just get it built.

The first thing is to create an opening for the door. In one corner of your 44¼ × 44¼-inch piece of plywood, mark out a 11³/₅ × 12¹⁵/₁₆-inch rectangle. The two sides of this opening nearest the corner should both be 1⅝ inches from the edges of the plywood. To cut out your rectangular opening, you can use your circular saw to do most of the work, but you might not be able to get all the way into your corners. Use a jigsaw or hand saw to finish. From this point, we're going to consider the corner with the hole to be the lower left corner of the plywood wall.

Next we'll put the trim on the door. Cut 45-degree angles on all ends of the two 14⅜-inch 1 × 1½-inch cedar trim pieces and the two 16-inch 1 × 1½-inch cedar trim pieces. Glue and nail these four pieces together to create a rectangular frame. Nail this into place over the hole you cut into the plywood.

BUILD INSTRUCTIONS (CONTINUED)

Before proceeding any further, we need to make a couple of cuts to modify the upper corners (opposite the door hole) of the plywood. As shown in the diagrams, make a mark 39½ inches up on both sides of the plywood wall. Then make two more marks on the top of the plywood, each 5⅝ inches in from their respective corners (leaving 33 inches between these two marks). Now make two cuts between each pair of 39½-inch and 5⅝-inch marks, thus removing the two upper corners from the plywood.

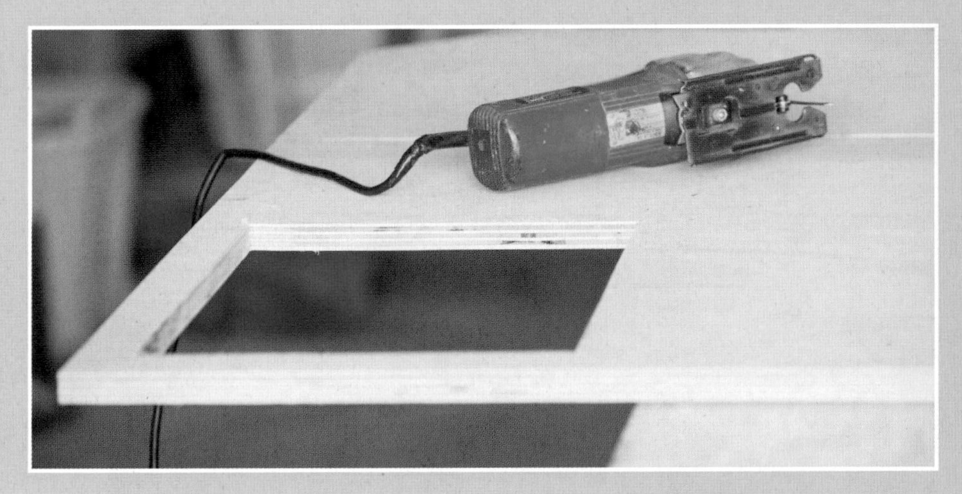

Top: A jigsaw can make the job of cutting the removable coop wall door very simple.

Middle: The finished doorway opening. Use a jigsaw to make precise cuts.

Bottom: Narrow pieces of wood can easily split accidently when they're being nailed, so take care. When used safely, an air nailer can be very good for handling this task.

Now you can shingle the plywood wall just as you shingled the other areas—but only the side with the door trim, of course.

There's one thing left to do, and that is to attach the door. Use the $12^{7}/_{8}$-inch piece of 1 × 12 cedar for the door, and attach with the pair of 2-inch hinges. Add a hook and eye latch, and you're finished.

Set your plywood wall aside, as it can't be installed until we have a final detail taken care of inside the coop—the nest boxes.

Top: Once you've completed your doorframe, attach it to the door opening you cut into the plywood.

Middle: Now you can shingle your removable coop wall to protect it and make it look great.

Bottom: Mount the coop's door onto the frame and add a hook to keep it closed as needed. You can also try other types of latches—some are even able to be opened and closed from the outside of the coop.

STEP 20: NEST BOXES

Cut list for the nest boxes:

- 1—33³/₈" × 13¹/₂" plywood, ¹/₂" thickness
- 2—33³/₈" × 3" plywood, ¹/₂" thickness
- 2—14" × 18¹/₄" plywood, ¹/₂" thickness

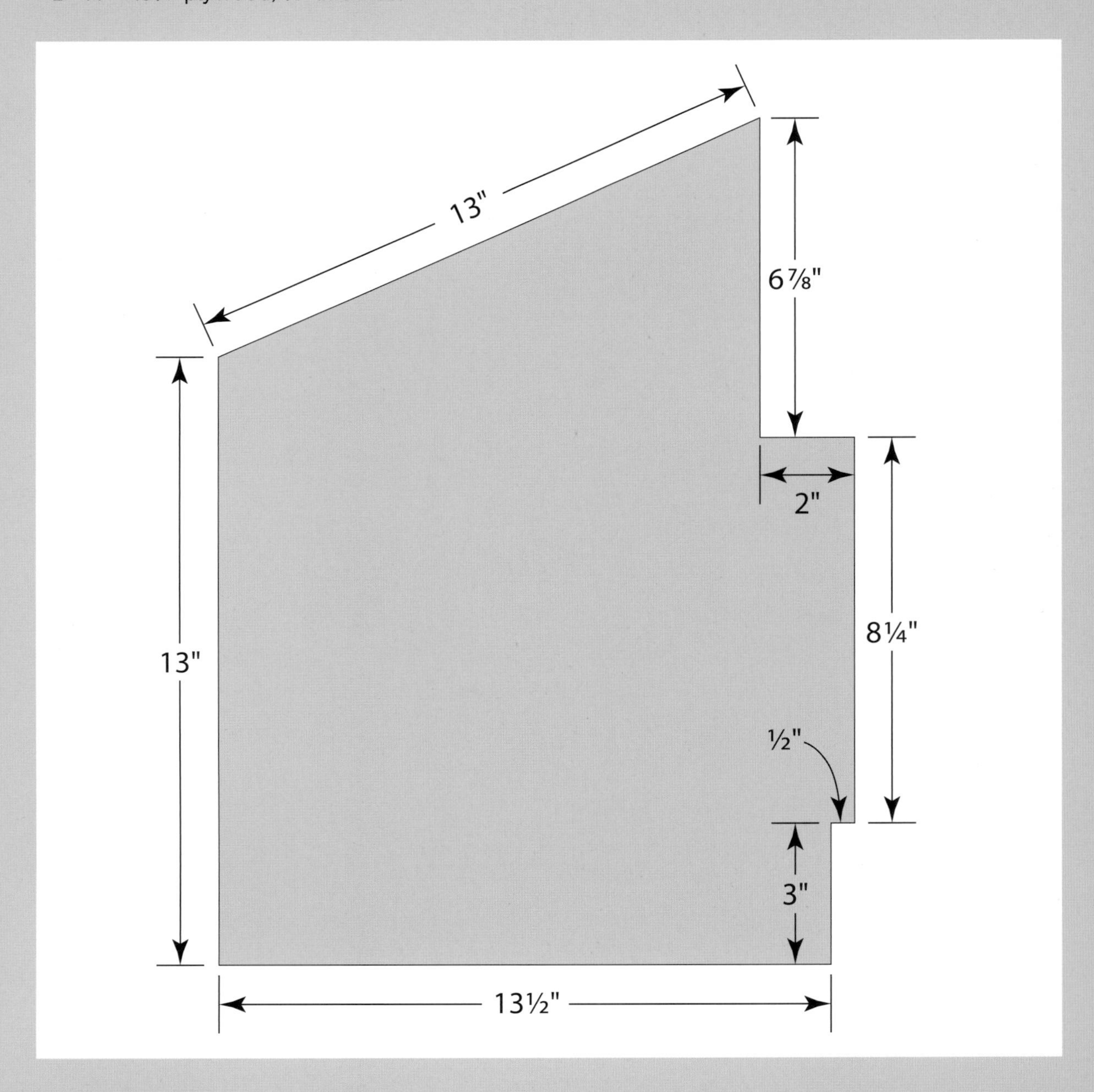

Building the nest boxes can be a little tricky, but they're very nice when you're finished. Following the diagrams carefully, make several adjustments to the two 14 × 18¼-inch plywood pieces—you'll be making two notches and slants for the nest box roof. The roof is made out of the 33⅜ × 13½-inch plywood, ripped to 65 degrees on one of the long sides. This would be a great time to borrow a jigsaw from someone if you don't own one.

The two 14 × 18¼-inch plywood pieces are spaced 10¾ inches from each other and from the outer edges of the roof as well.

The two 33⅜ × 3-inch plywood strips are used as a base to keep the nest boxes sturdy. Nail all of the sections together.

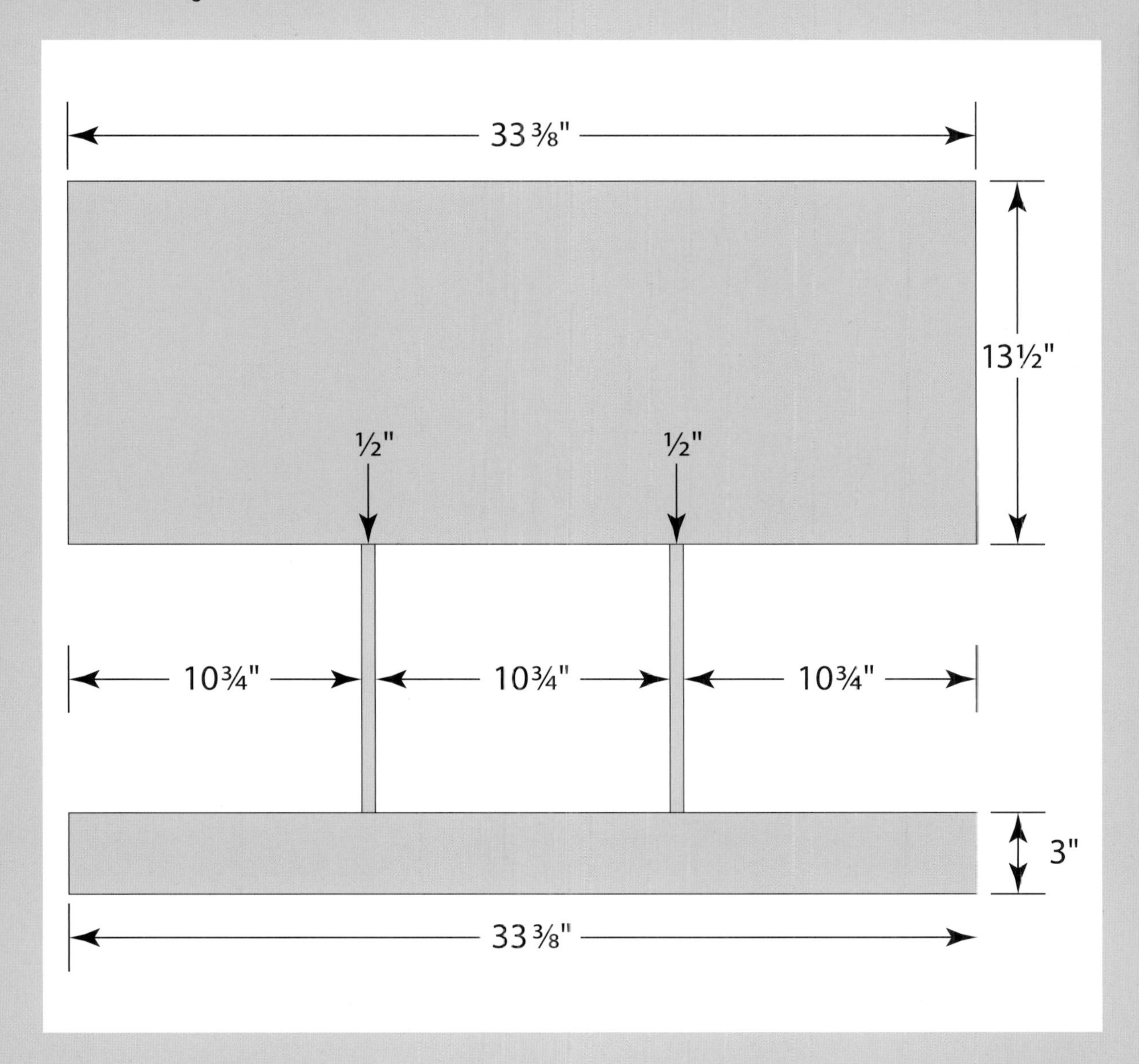

BUILD INSTRUCTIONS (CONTINUED)

One final detail is to make two minor cuts to the rear $33^3/_8$-inch plywood strip (the one that was placed in the slots). As shown, remove $^3/_8$ inch from one end (nearest the back wall once the nest box is installed) and $2^5/_8$ inches from the other end (the one that will be closer to the door).

You can anchor your nest box to the coop if you wish; use screws so the nest box can be removed for cleaning if desired.

Assembling the nest boxes is fairly straightforward, but you might need an extra pair of hands to hold the pieces in place as you nail.

Take your nest boxes into the coop to mount along the wall. If you prefer, you could make your nest boxes larger and have two instead of three.

This is what the interior of the coop looks like before the final interior wall is in place, showing the position of the nest boxes and the roosts.

Now you can fasten the final, removable coop wall with 3-inch screws.

STEP 21: INSTALL THE REMOVABLE COOP WALL

Next you can attach the coop wall with 3-inch screws. Things are really looking good now!

This is the interior of the coop with the final wall in place, to show its location.

STEP 22: BUILDING THE CHICKEN LADDER

Cut list for the chicken ladder:

- 1—54" 1 × 12 cedar or pine
- 13—½" × ¾" × 11¼" cedar slats

Hardware:

- 2—hook latches and eyes

Since the actual coop portion of your coop is raised off the ground, your chickens need a way to climb up there and access it. A simple ramp made of the 54-inch 1 × 12 will do the job. To make it easier for the chickens to climb, install thirteen ½ × ¾ × 11¼-inch slats, spaced 3¼-inch apart, along the 1 × 12 to provide your chickens with steps. Attach the slats with nails.

The chicken ladder needs two eye hooks at the top to allow it to be hung in place inside the coop.

Now it's time to install the chicken ladder. The slats provide chickens with a way to grip the ladder and climb up.

STEP 23: NEXT BOX AND CLEANOUT DOORS

Cut list for the nest box and cleanout doors:

- 2—30¼" 1 × 12 cedar

Hardware:

- 4—2½" hinges
- 2—latches

The egg collection door allows access to the nest boxes without having to go inside the coop.

Guess what? You're almost finished constructing your coop! The next step is to install doors over the exterior access openings on either side of the coop—one covering the nest boxes, to allow for easy egg collecting, and another on the opposite side, to allow for easy cleaning of the coop.

To create the doors (which are identical), place one 30¼-inch cedar 1 × 12 in each of the openings on the sides of the coop, then secure each one in place with a pair of 2½-inch hinges. The hinges should be placed on the bottom side of the cedar 1 × 12s, so that the doors open downward, not upward or to the side. Attach a latch at the top of each door so that they will remain shut when not in use.

The cleanout door is identical to the egg collecting door and allows you to clean the inside of the coop from the outside. Handy!

STEP 24: DOORSTOP AND FINAL HARDWARE CLOTH

Cut list for the doorstop:

- 1—34¼" 1 × 4 cedar
- 2—62" 1 × 4 cedar

To make your coop door work really well, install a doorstop around the doorframe. Use the two 62-inch 1 × 4s on either side of the door and the 34¼-inch 1 × 4 for the top. Let all three of the 1 × 4s overhang about a half inch so that the door will bump up against them and not be able to swing inward into the coop.

There are a couple of final places where you need to attach hardware cloth: on the two narrow openings on either side of the door and in the gable opening above the door. Attach the hardware cloth with roofing nails.

A simple doorstop just inside the coop will prevent the door from swinging too far inside.

Just a couple of final details and your coop will be complete. You'll need one more extra piece of hardware cloth over the door . . .

. . . and two strips of hardware cloth to cover the additional areas on either side of the door.

STEP 25: FINISHING THE COOP

Coops made of cedar (like this one) can be left unfinished; cedar doesn't need paint or stain to protect it from the elements the way other types of wood do. However, you can certainly apply a coat of exterior paint or stain to a cedar coop for decorative purposes if you wish.

And now you're done! Move in your chickens and enjoy the coop!

Finished! The coop is now ready for your chickens to move in. You should be proud.

The finished coop from the back.

Feel free to add your own personal decorating touches.

6 ADAPTING AND CONVERTING COOPS

What if the coop in Chapter 5 doesn't quite meet your needs? Maybe your coop needs to be a bit larger—large enough to house, say, ten birds instead of four. Or maybe you need a coop that is easily portable, or one that offers a larger run.

In this chapter, we'll show you how to adapt the basic coop plan to suit your needs. We'll also introduce you to ways that you can convert existing structures into chicken coops potentially saving time and money.

ADAPTING THE BASIC PLAN TO CREATE A LARGER COOP

Perhaps you aspire to keep a larger number of chickens than the coop in this book is designed for. If this is the case, you might want to construct a larger coop, without creating a design totally from scratch. What can be done?

First off, you could lengthen the side panels. Instead of using 96-inch (8-foot) 2 × 4s for the side panels, consider extending them

Above: If you have aspirations to someday move on to a larger flock, you may want to consider modifying the coop plan in this book to accommodate more birds.

Opposite: Now that the Chapter 5 coop is complete, it's time to move in those adorable chicks and hens. But maybe you're looking to add more room by converting another, existing structure into a chicken coop. In this chapter, we'll take a look at ideas for adapting and converting coops.

another 29¼ inches using 11-foot 2 × 4s cut to 125¼ inches. This will give you enough length to add an additional section to the run. You'll also need one additional rafter, longer (125¼ inches) 2 × 12s for the roof, and additional shingles.

If you'd like to make the coop area larger, don't worry—it can be lengthened as well. Increasing the coop's interior length from 30⅜ inches to 44⅝ inches will give you enough room for an additional nest box, not to mention more elbow room (or maybe it's wing room?) for more chickens. You could even choose to go longer—all the way to the next 2 × 4 stud. Of course you'll also need to lengthen the cleanout and nest box doors, as well as the coop's floor, but these modifications are relatively straightforward and can actually make for a fun challenge.

The advantage to lengthening these areas—rather than making them wider—is that the rafter, door, and removable coop wall measurements remain intact and don't have to be reconfigured.

ADAPTING THE BASIC PLAN FOR WINTER

Have you ever heard anyone who keeps livestock say, "I just love winter, it's my favorite season"? Chances are, the answer is no. Winter can be a challenge for chicken keepers who live in cold areas. But if you live in a particularly cold region with hard winters, you're in great shape, because this coop has already been designed with cold weather in mind. There is excellent ventilation without letting in drafts, while the thick 2 × 10s and 2 × 12s that are used on the walls and roof offer fine insulation properties.

Nevertheless, there are a few more things you can do to make your coop better equipped to handle cold weather.

In addition to building your coop with cold weather in mind, you might want to consider a chicken breed that is naturally hardy in winter, such as the Russian Orloff.

Even though this coop is suited well for winter housing, it's a good idea to use winterization panels on the run areas of any coop, as discussed in Chapter 3. To winterize the run with wooden winterization panels, you'll need to install a frame of 1 × 1-inch spacers around the outside of each area of hardware cloth. Then use ½-inch plywood to create the panels. Wing nuts can then be used to hold the panels in place.

WINTERIZATION PANELS

Removable winterization panels are fairly easy to make and can go a long way toward keeping your chickens warm all winter. You can build winterization panels out of Plexiglas (which is a flexible, clear, durable plastic) or wood.

Both Plexiglas and wood winterization panels can be built to slide into place on any areas that are normally covered with hardware cloth and therefore open to the air. You can protect open areas of both the coop and the run to provide insulation from the cold without sacrificing the beauty of your structure.

In general, try to use Plexiglas panels on the sides of your coop where the sun is likely to shine; this will depend on trees and other objects near your coop, but it's likely you'll want to install Plexiglas on the southern side of your coop. Then use your sturdier wooden winterization panels to provide protection against the direction of the prevailing winter winds.

HEATED WATER

If you've raised any kind of animal before, you are aware that providing water during winter can be a challenge. If your chickens cannot access their water, you're in trouble. When the temperatures consistently drop below freezing, consider an electric, heated water pan, which will keep your chickens' water from freezing even in very cold conditions. You'll need an electrical source, of course, as described earlier.

ADAPTING THE BASIC PLAN TO CREATE A PORTABLE COOP

You may like the idea of a coop that is semi-portable. To be sure, a chicken coop of any variety (except perhaps a chicken tractor) isn't something you're going to want to move around every day. But it isn't a bad idea to move your coop throughout the year to take advantage of new grass or to choose a better location. Let's look at a couple of ways you could adapt the basic coop plan to make it easier to move.

If you want your coop to be portable, you need to make it lighter. One way to decrease the weight of your coop is to side the exterior walls with plywood siding such as T1-11 (sometimes mistakenly called bead board) instead of shingles. Doing this saves a lot of weight because you can eliminate the heavy 2 × 10s that make up the coop walls. T1-11 is a flimsier product, however, so you may need to consider incorporating additional 2 × 4 studs to the back and side panels to help strengthen the T1-11 and give it something more to hold on to. For instance, an additional 59¾-inch 2 × 4 could be added vertically to the center of the back panel to provide additional strength. Likewise, additional 30⅝-inch 2 × 4s could be added horizontally to the side panel areas. T1-11 comes in 4 × 8-foot sheets, so by purchasing just three sheets you should have more than enough

T1-11 is another option for the walls of your coop. It weighs less, but needs to be painted for protection from the elements.

material to cover the back wall, side walls, and removable front wall of your coop.

You could even go as far as eliminating siding altogether on the portions of the side panel walls beneath the coop and instead opting to close these off with hardware cloth.

It's possible to make the roof lighter (although perhaps not quite as cute) as well. The six cedar 2 × 12s that make up the base of the roof could be replaced with ¾-inch plywood and then covered over with ordinary roofing shingles, or even a metal roof attached with screws.

Of course, even with these modifications, your coop is still going to be rather heavy. Therefore, a set of wheels will go a long way to making your modified coop truly portable.

You don't want to have your coop raised up on wheels like a cart, since this would give predators easy access. Instead (as shown in the photo at right), consider installing a pair of wheels on just one end of the coop by taking those base 96-inch 2 × 4s on the side panels and extending them another 12 inches to 108 inches. Your wheels can be attached to these extensions, and corner braces can be made out of additional 2 × 4s. Using this method, the wheels will stick flat out the back, letting the coop stay flush on the ground. The wheels only come into play when it's time to move the coop—and then they act like wheelbarrow wheels. The idea is that you raise up the end of the coop opposite the wheels (perhaps by installing handles or with the help of a machine) and let the wheels carry most of the load. For this reason, you'll want to install the pair of wheels under the coop end of your structure. Remember, these are guidelines, and your actual modification decisions will depend on the materials that are available to you—and your skill level.

ADAPTING THE BASIC PLAN TO CREATE A DIVIDED COOP AND/OR BROODING AREA

You may find it useful at times to be able to divide your coop into two areas. Luckily, modifications can be made to the design in Chapter 5 that allow for this. One option is to isolate the area directly under the coop and turn it into a divided area. Simply use chicken wire or hardware cloth to separate the area under the coop from the rest of the run (this way the wire is semi-removable and will allow you access when needed). An outdoor access door for the chickens could be either framed out with 2 × 4s, or even installed on the back wall of the coop by cutting through the 2 × 10s.

Extending the coop's base by a few inches and adding braces and wheels can help turn your coop into a somewhat more portable version.

CONVERTING OTHER STRUCTURES

Building a coop from scratch is a rewarding experience, but it does require an investment of time and energy. You can shave some time off your coop project by renovating an existing structure. Maybe you have an old playhouse that is no longer being used for childhood tea parties and you'd like to give the structure a second life. Or maybe you have a garden shed with a lot of life left in it, or a doghouse with definite coop potential.

In the following pages, we'll explore options for renovating and converting existing structures and turning them into chicken-friendly habitats. We'll discuss necessary considerations, benefits, drawbacks, and additional information that will help you decide whether to build brand-new or convert an existing structure into the coop of your dreams.

But don't limit yourself to these ideas. Opportunities abound when it comes to repurposing coops. If you need more ideas than we've outlined here, simply search online for "converting structures into chicken coops." Or stop by Pinterest to view an innumerable assortment of impressively clever suggestions for repurposing buildings into chicken coops. You'll undoubtedly find inspiration to convert a chicken coop of your own.

ADVANTAGES OF CONVERSION

Renovation can be an effective way to build a chicken coop without cost playing a major role. Another reason to convert an existing building into a chicken coop is that it can be much more fun! Being faced with many different layout styles is sometimes daunting, so renovation can be a friendly approach to creating a coop. The

Do you have an unused building that could have a second life as a chicken coop? The possibilities exist. This old log sauna was converted into a chicken coop by cutting windows and a chicken door and adding an outdoor run.

If you're thinking about using an existing building to gain more room, you could think about adding extra nest boxes for your growing flock.

ability to recycle is another reason to update an older structure. Instead of just letting unused outbuildings be wasted, repurpose them into something you'll need every day. Converting is particularly appealing for those who are not as familiar with construction.

Choosing what to convert is one of the hardest challenges, but you might have several options already. Many farms have outbuildings that could be converted into delightful coops. Former doghouses, rabbit hutches, children's playhouses, and storage sheds are just some of the buildings you can repurpose into lovely chicken coops. Do you have a couple of empty box stalls that aren't housing horses? Convert them into splendid large-space chicken coops.

When you have a building that could stand to be repurposed, you might be more open to getting creative and taking a less standard method with your chicken coop. The personalization process can involve the entire family. Choosing a paint color and doing the actual painting is something even younger family members might be willing to assist with. You won't have to purchase as many materials as you would when starting from scratch, so you'll be able to spend more time beautifying the coop. Remember, safety is the number one concern when renovating a coop, so be sure that you keep this in mind while working.

DISADVANTAGES OF CONVERSION

What if you have your heart set on a 6 × 10-foot chicken coop and the shed that you'd like to renovate is only 4 × 8? The lack of the ability to customize your coop can be a disadvantage of coop conversion, but it's not an insurmountable problem. It's wise to look at your existing structure as a baseline, but not necessarily the final blueprint of your renovated coop.

POSSIBLE BUILDINGS TO CONVERT

If you live on a farm, there's a good chance that your property has a number of small outbuildings that are perfect to convert into a chicken coop. But even if you're not a farmer, there are plenty of possibilities.

SHEDS

It often seems like the farmers of the olden days couldn't get along without one or two small sheds, perfect for storing miscellaneous equipment and whatnot. If you don't live on a farm, you might still have a shed out in the corner of your yard, harboring some of Dad's less frequently used tools or a collection of rusting garden supplies. If you have the good fortune to possess such a shed, and if its present

HEALTH CONCERNS

When converting an older structure into a chicken coop, keep in mind the potential health hazards involved. While any construction project has its hazards, there are a few special concerns that come with older structures.

One obvious concern is encountering old, rusty nails. Chances are that any older structure will contain a few (or maybe more than a few!) old nails. As nails can cause serious injury, it is wise to be extremely cautious when dealing with them. If you will be removing old nails, or wooden boards containing old nails, be sure to put each nail in a bucket so it cannot fall to the ground where it could be stepped on. Furthermore, wear thick clothing to protect yourself from being scratched by a nail that you didn't notice.

Another major concern is that the building you are converting may not be 100 percent stable. If your building seems rickety, unbalanced, or severely deteriorated in some manner, it may be in danger of collapsing. Obviously, you don't want to be inside any building where collapse is a possibility, so try to confirm the soundness of the building before you begin converting it into a coop.

You will also want to keep your eyes open for lead paint. As you may know, many old paints were manufactured with lead. Unfortunately, lead can be poisonous if ingested or inhaled by you or your chickens. Thus, if the building you're interested in converting has an old layer of paint, you may wish to have a professional remove the paint from your building, or choose a different structure for conversion.

With the right adjustments, a simple garden shed can become a chicken coop.

Some cute decorating, some windows, an outdoor run, and this shed is now a chicken coop.

A chicken coop doesn't necessarily need to be its own building. If there is spare room inside a larger barn, you could consider converting a portion of the building into a coop.

purpose is indeed to house various objects for infrequent use, then by all means bid farewell to your hodgepodge shed and say hello to your new chicken coop.

The main thing to consider when converting a shed into a chicken coop is that many sheds lack any form of ventilation, having been built for the sole purpose of storing tools, equipment, and other objects that don't require a constant supply of fresh air. Therefore, think about cutting big windows in the sides of the shed to supply much-needed ventilation and light. In some cases, this may be as easy as cutting openings into the walls and installing the necessary components; in other instances, it may require major structural changes to the shed that should only be undertaken by a professional carpenter.

One great advantage to adapting a shed is the extra size you have to work with. If you were to convert an entire basic 8 × 10-foot shed into a coop (and have an outdoor run attached separately), it could easily house as many as twenty chickens. Or you could just wall off the back end of the shed for use as the coop and possibly remove siding up front to create a built-in run. There will be plenty of room to stagger your roosts in the corners, as well as no lack of space for nest boxes and other amenities.

BARNS

Although a big old barn for livestock isn't quite what people tend to picture when thinking of a place to keep chickens, there can be no denying that a traditional livestock barn is a wonderful option for protecting your poultry. While it may not be as picturesque and quaint as a smaller, more typical chicken coop, a barn offers several major advantages that smaller buildings can't provide. For example, you can easily convert a small portion of a barn into a top-notch chicken coop while still retaining the vast majority of the barn for other purposes, a major advantage if you don't have room for a standalone chicken coop or if your farming endeavors extends to livestock other than chickens. Also, barns typically have electricity installed for running lights and heaters, so powering the chicken coop portion of your barn will not be an issue—yippee!

However, like sheds, barns often require a bit of work in the ventilation department. For while barns do have windows, the amount of air movement in the portion of the barn where your coop is located might not offer enough fresh air for a flock of chickens. Adding windows or vents could be a challenge, so in the case of a barn—which is most likely already equipped with general ventilation—some well-placed electric fans to circulate fresh air may be all you need to keep your chickens happy and healthy. You might need a chicken-sized door to allow them to access the outdoors and perhaps a run attached to the outside wall of the barn.

CHILDREN'S PLAYHOUSES

You may have many fond memories of pleasant afternoons spent playing with stuffed animals

Repurpose an old playhouse for new tenants.
© Getty Images

and board games in the playhouse your dad built for you, but if the playhouse isn't getting as much use as it did in its heyday, perhaps the time has come to give it a new lease on life by converting it into a chicken coop. In general, playhouses are the perfect size to be adapted into chicken coops, and in some cases they already have electricity installed. They also might already have good ventilation, or you may still need to make adjustments in this department. There will probably already be enough windows—they'll just need hardware cloth installed to protect against predators. But all in all you probably won't have to do much to prepare your playhouse for chickens. Really, outside of installing nest boxes, roosts, and such, many playhouses barely need any converting at all.

LEAN-TOS

Like sheds and barns, lean-tos can easily be adapted into exceptional chicken coops. Perhaps you have a lean-to that was once used for storing wood, back before your house was heated with gas or electricity instead of a wood-burning furnace. If your lean-to is just sitting there gathering dust and seemingly begging to be put back into good use, you may have

Converting different kinds of buildings into coops can sometimes be as easy as cutting doors and adding ventilation—and the all-important chicken ladder.

While usually quite small, a converted doghouse has the potential to house a couple of chickens.

found yourself the foundation for your new chicken coop.

The open side of a lean-to can make a great opportunity for constructing an attached run. Of course, this also means that you'll have to put up an additional wall to give your chickens protection from the elements as needed, but this shouldn't prove too much of an issue.

As is becoming a bit of a theme here, your primary concern is to make sure that your lean-to offers good ventilation. Depending on how you construct your coop, you may not have to add much in the way of windows and/or vents. However, if you deem it necessary to add additional ventilation, the simplistic design of lean-tos should allow you to easily cut sizeable openings in the walls, in which screened windows or vents can be installed.

DOGHOUSES

You know that doghouse out in your yard—the one being engulfed by tall grass and weeds? The one you built when Fido was a puppy, back before he started sleeping in your bed at night? Even if it was never used for its intended purpose of housing your canine companion, a good-sized doghouse can easily be adapted into top-notch living quarters for your chickens.

Right: At the end of the project, you'll have the satisfaction of knowing that you built your own coop with exactly the features you desired. It's a great feeling—now go have fun with your flock!

Below: Keeping chickens is a hobby that can involve the entire family—even the youngest members. Don't let them miss out on the fun!

Of course, unless your doghouse was built for a great dane or a Mastiff—or unless you intended to spoil Fido with a doghouse of epic proportions—chances are your doghouse isn't very big. Therefore, you may be limited in the number of chickens you can keep in a coop converted from a doghouse. If you weren't intending to keep more than two or three, this won't be a problem—your coop will be just the right size. But if your chicken-raising dreams involve a larger flock, you might be better off building a new, large coop rather than trying to adapt a doghouse to suit your needs.

As most doghouses rarely (if ever!) have windows or vents, you'll need to add some to provide your chickens with proper ventilation. A good place to do this is in the gables at each end of the doghouse, assuming it has a gable roof.

TIPS FOR MAKING CHORES EASIER

During the first few weeks and months of your chicken-raising experience, the time spent

Chore time! Some chores are enjoyable and fun, some not so much. Either way, there are usually ways to speed up repetitive tasks and make life simpler. Why not plan ahead and free up more time to enjoy your hens?

You can simplify your morning chores (and eliminate the possibility of forgetting to close up at night) by using an automatic coop door that operates on a timer. This one is solar-powered.

You might not immediately think to add a thermometer to the inside of your coop, but it's a good idea. It will help you easily and quickly monitor temperatures that may require action to keep your birds comfortable.

performing chicken-related chores may seem insignificant—especially if you only have a few birds in your flock. But over time, small inefficiencies in your chore workflow can add up to a significant loss of productivity, so it's often worth the effort to carefully evaluate ways that you can save a few minutes here and there. Naturally, there will be many aspects of chicken keeping that you enjoy—collecting eggs, watching the flock interact—but there will likely be some chores that you'd prefer to accomplish as quickly as possible. Here are a few ways you can increase the efficiency of your work.

INSTALL A SOLAR-POWERED DOOR TIMER

Let's assume your chickens have access to a larger run via a chicken door leading from the main coop, as shown in the photo at left. Your hens will likely return automatically to the coop at dusk after a long day of pecking and frolicking. But what *won't* be automatic is the closing of the coop door. Someone must go out each evening to close the door, then return in the morning to open it up. This isn't a very time-consuming task, but it *is* a chore that occurs over and over and over. This type of chore is an excellent candidate for automation, and that's exactly what the chicken owner in this photo has done. A simple solar panel mounted on the roof of the coop supplies electricity for a timer, which opens and closes a small chicken-sized coop door at the appropriate times. A machine like this not only saves time, but also acts as a guard against human forgetfulness and helps out on early summer mornings—when it's light out at six o'clock and the chickens are ready to go out, but you're ready to stay in a bed awhile longer. A timer can let them out all by itself!

WORK FROM THE OUTSIDE IN

No matter what kind of livestock you're working with—goats, cattle, rabbits, chickens—it's almost always easier and faster if you can do basic jobs like feeding, watering, and opening/closing doors *without having to go in with the animal.* For chickens especially, being able to perform simple jobs from the outside of the coop can speed you up considerably, since it cuts down on time spent changing into your work footwear, trying to open the door without letting the hens out, walking through/around them, and so on.

The plans for the main coop project in this book incorporate a couple of time-saving designs, including a door to the nest boxes that can be accessed from the outside and an external coop cleanout door. Other time-saving options are limited only by your ingenuity.

Why go inside the coop to open the interior door? With a simple extended handle . . .

. . . you can do it from the outside.

When dealing with poultry, you save time whenever you can avoid actually going in the coop to perform chores such as egg collecting.

Basic waterers and feeders can be created out of PVC pipe; they're a simple and easy DIY project.

In the photos at left, notice how the interior coop door can be opened and closed from the outside via a rod that connects to the coop door.

USE AUTOMATIC WATERERS/FEEDERS

While you may enjoy the process of feeding and watering your flock each day, don't overlook the possibility of incorporating automatic feeders and waterers into your coop. You can find sophisticated versions at farm supply stores, but you can also build yourself simple-yet-effective "automatic" feeders and waterers that are completely gravity-fed and require no electricity or moving parts at all. There are multiple designs for this type of setup, but the general idea centers on a large storage reservoir of feed or water that is inaccessible to the chickens. The reservoir delivers its goods to a simple feeding or watering system that constantly resupplies itself as the chickens eat or drink. In this way, it is possible to feed a large number of chickens quite quickly, or even ensure that they will have enough food and water for more than one day if needed.

Remember to enjoy quiet times with your chickens. Sometimes the most enjoyable memories can occur during the smallest moments.

Showing your chickens at a local fair is a fun experience, and a great way to participate in an event with other nearby poultry enthusiasts. You'll have the chance to enjoy the competition, and perhaps even come home with a prize!

That's it! If you need more help with building ideas or caring for your chickens, check out the list of resources at the back of this book. Good luck, and have fun!

RESOURCES

SUPPLIES

CEDAR CHICKEN COOP.COM
cedarchickencoop@gmail.com
cedarchickencoop.com
www.facebook.com/Cedarchickencoopcom
920-420-5550

CHICKENCOOPGUIDES.COM
4283 Express Lane
Suite 233-740
Sarasota, FL 34249
www.chickencoopguides.com

FARMTEK
1440 Field of Dreams Way
Dyersville, IA 52040
www.farmtek.com
www.facebook.com/FarmTek
800-245-9881 (customer service)
800-327-6835 (sales)
563-875-2288 (retail store)

GQF MANUFACTURING COMPANY
2343 Louisville Road
Savannah, GA 31415-1619
sales@gqfmfg.com
www.gqfmfg.com
912-236-0651

JEFFERS
P.O. Box 100
Dothan, AL 36302
www.jefferspet.com
www.facebook.com/JeffersPet
800-533-3377

KEMP'S KOOPS INCUBATORS
sales@poultrysupply.com
www.poultrysupply.com
www.facebook.com/kempskoopsincubators
888-901-2473

NASCO—FORT ATKINSON
P.O. Box 901
901 Janesville Avenue
Fort Atkinson, WI 53538-0901
custserv@eNasco.com
www.eNasco.com
800-558-9595

PREMIER 1 SUPPLIES, LTD.
2031 300th Street
Washington, IA 52353
www.premier1supplies.com
www.facebook.com/premier1supplies
800-282-6631
319-653-7622

SMITH POULTRY AND GAME BIRD SUPPLY
14000 West 215th Street
Bucyrus, KS 66013-9519
smithkct@centurylink.net
www.poultrysupplies.com
913-879-2587

SNAP LOCK CHICKEN COOPS
601 Hurricane Shoals Road, NW
Lawrenceville, GA 30046
www.snaplockchickencoops.com
800-310-3867

TRACTOR SUPPLY COMPANY
www.tractorsupply.com
www.facebook.com/TractorSupplyCo
877-718-6750

© Lantapix / Shutterstock

HATCHERIES

CACKLE HATCHERY
P.O. Box 529
Lebanon, MO 65536
www.cacklehatchery.com
www.facebook.com/CackleHatcheryMO
417-532-4581

ESTES FARM HATCHERY LLC
11299 Lawrence 1163
Mt. Vernon, MO 65712
www.esteshatchery.com
800-345-1420

HOOVER'S HATCHERY
P.O. Box 200
Rudd, IA 50471
sales@hoovershatchery.com
www.hoovershatchery.com
www.facebook.com/HooversHatchery
800-247-7014

IDEAL POULTRY BREEDING FARMS, INC.
215 West Main Street
Cameron, TX 76520
sales@idealpoultry.com
www.idealpoultry.com
www.facebook.com/idealpoultry
254-697-6677

MEYER HATCHERY
626 State Route 89
Polk, OH 44866
info@meyerhatchery.com
www.meyerhatchery.com
www.facebook.com/meyerhatchery
888-568-9755
419-945-2651

MOUNT HEALTHY HATCHERIES, INC.
9839 Winton Road
Mt. Healthy, OH 45231
info@mthealthy.com
www.mthealthy.com
www.facebook.com/mthealthyhatcheries
513-521-6900

MURRAY MCMURRAY HATCHERY
P.O. Box 458
191 Closz Drive
Webster City, IA 50595
www.mcmurrayhatchery.com
www.facebook.com/MurrayMcMurrayHatchery
800-456-3280
515-832-3280

MY PET CHICKEN
info@mypetchicken.com
www.mypetchicken.com
www.facebook.com/pages/My-Pet-Chicken/149928893560
888-460-1529
908-795-1007

STROMBERG'S CHICKS AND GAME BIRDS
100 York Street
P.O. Box 400
Pine River, MN 56474
info@strombergschickens.com
www.strombergschickens.com
www.facebook.com/StrombergsChickens
800-720-1134
218-587-2222

TOWNLINE POULTRY FARM, INC.
P.O. Box 108
Zeeland, MI 49464
office@townlinehatchery.com
www.townlinehatchery.com
www.facebook.com/townlinehatchery
888-685-0040
616-772-6514

© Cynthia Farmer / Shutterstock

INDEX

ACKNOWLEDGMENTS

Writing a book is always an enormous undertaking, and we're endlessly indebted to the many people who share their time, talent, expertise, and encouragement with us as we navigate the book from start to finish. We owe an extra special thanks to the following individuals:

- **QUARTO PUBLISHING GROUP, VOYAGEUR PRESS**, our editors Elizabeth Noll and Jordan Wiklund, and everyone who helped us bring this book to completion—thank you!
- **LORIN**, for sharing his construction knowledge and helping craft the project coop.
- **PAULETTE**, for photo editing, photo shoot coordination, and endless help.
- **KEELER, EMILY, AND ANNA**—for everything. You guys are completely awesome!
- **GREG AND MARIE RIPPEL**, for sharing their time, expertise, and beautiful coop and chickens with us.
- **GINA GOLLIHER AND FAMILY**, for sharing your lovely chickens and coops with us.
- **PEACHES**, because she wants to be mentioned.
- **GRACIE**, because she wants to be mentioned too.

ABOUT THE AUTHORS

As brother and sister collaborators, **DANIEL AND SAMANTHA JOHNSON** pursue their writing, photography, and agricultural interests at the family-owned Fox Hill and Pine Valley Farms in northern Wisconsin. Together, they have cowritten and photographed several books, including *Horse Breeds: 65 Horse, Draft, and Pony Breeds*; *How to Raise Horses*; *How to Raise Rabbits*; *The Beginner's Guide to Vegetable Gardening*; and *How to Build Chicken Coops*. Since 1999, they have been involved with raising and showing registered Welsh mountain ponies, and they also keep an assortment of purebred rabbits, including mini rexes and Holland lops. Several hundred thousand honey bees also make their home at Fox Hill and Pine Valley.

DANIEL JOHNSON, writer and photographer, likes to spend his time lugging around heavy camera equipment in all kinds of weather to take pictures of things such as dogs pulling sleds at 20 degrees below zero or people hauling hay in 90-plus-degree heat. Dan films and produces videos for companies to use on websites and in online magazines. In his spare time, he also photographs frogs, one of which has been his pet for the last twenty-two years.

SAMANTHA JOHNSON is an award-winning writer, as well as a proofreader and pony wrangler, gardener, and farmer. She is also a horse show judge, is certified with the Wisconsin State Horse Council and the Welsh Pony and Cob Society of America, and has judged horse shows across the United States from Maryland to California and locations in between. Samantha enjoys making to-do lists, watching old episodes of *Little House on the Prairie*, and daydreaming about buying a couple of Cheviot sheep and a miniature Jersey cow.